D0626947

WITHDRAWN

Molecules and Minds

303.483
R72m

STEVEN ROSE

Molecules and Minds:

Essays on biology and the social order

Open University Press

Milton Keynes • Philadelphia

Open University Press
Open University Educational Enterprises Ltd
12 Cofferidge Close
Stony Stratford
Milton Keynes MK11 1BY, England
and
242 Cherry Street, Philadelphia,
PA 19106, U.S.A.

First published 1987

Copyright © 1987 Steven Rose

All rights reserved. No part of this publication may be reproduced, stored in a retrieval system or transmitted in any form or by any means, without written permission from the publisher.

British Library Cataloguing in Publication Data

Rose, Steven
Molecules and minds : biology and the social order.
1. Biology —— Research —— Social aspects
I. Title
303.4′83 QH333

ISBN 0-335-15814-5

ISBN 0-335-15813-7 Pbk

Project Management: Clarke Williams
Printed in Great Britain

Cover illustration: Raster display of tomato bushy stunt virus, by Nelson Max and Arthur Olson (Lawrence Livermore National Laboratory, Livermore, California). Reproduced with permission.

Contents

88-1932
ALLEGHENY COLLEGE LIBRARY

Introduction

For my eighth birthday I was given two presents I'd specially asked for — a chemistry set and a copy of *The Origin of Species*. I wasn't as precocious as perhaps I'd hoped, and it was some years before I got beyond the first few pages of *The Origin*, but it did give me the intellectual weaponry I needed for those fierce whispered debates that took place every Saturday morning among the young boys at the back during the more boring bits of the interminably long synagogue service. Where did we come from then, if god didn't make us — didn't even exist? Monkeys and ultimately amoeba, I could answer firmly. Where did they come from? And where did the world come from? Sucked out of the sun in a giant collision with another star would be the answer (it was the days before Big Bang theory of course). With the power of The Book, and a lab in a garden shed, filled with copper sulphate, iron filings and other exotica earned with pennies gained by washing bottles for the local chemist, I was well set up; I was going to be a scientist, that was clear.

But my home was also full of politics. Back from the Second World War, my father had become for a time full-time anti-fascist organiser for the Association of Jewish Ex-Servicemen — there was a surprising amount of fascism and anti-semitism about in London in the late 1940s, despite the fact that ostensibly the war had been fought to destroy it. There was anti-semitism at my school, one of the old London guild schools that cared sufficiently for its reputation to operate a 10% quota system on Jewish boys wanting to enter, and instructed its selected pupils in the subtler details of the British class system by lecturing them against mixing with the *hoi-polloi* from the Council school just down the road. (Odd, in retrospect, that one of my fellow ethnic minority at that school was, in the 1980s, to become a Tory Home Secretary entrusted with enforcing, with every appearance of commitment, an exclusion policy on immigrants which would certainly have prevented his own family, like my own grandparents, settling in Britain at all.)

When I entered university as an aspirant chemist — soon to become a biochemist — my scientific and political concerns ran is parallel, without convergence. I took it for granted that science and politics didn't mix; rather, science was part of the world of rationality towards which political action should strive. Like an older generation of left scientists (as I know

now but didn't then), it seemed clear to me that science was on the side of socialism; that a rational and beautiful world, one which by right political action we could bring to pass, was also one in which the scientific world-view would prevail.

I wanted to do what I could to bring that about and in the meantime get on with my research. You couldn't be a biochemistry student in Cambridge in the late 1950s without being captured by the excitement of the new biology. History was being made. In my first undergraduate year Sanger got the Nobel Prize for determining the chemical sequence of the insulin molecule and the foyer of the Department was awash with champagne. Down the road, as everyone knew, Crick and Watson had just solved the structure of DNA (no one spoke about Rosalind Franklin then). Biochemistry was the key to understanding the world and it would bring together in due course a universal synthesis, of the theory within *The Origin of Species* and the games I played with my chemistry set. And this was no mere prospectus for some remote future; it was happening, it seemed, here and now, in the laboratories around me. Asked what research I would do when I graduated, I replied with the arrogance of youth. All the interesting DNA stuff had been done. The task of biology now was to solve problems about the brain. It took molecular biologists about another fifteen years to come to the same conclusion — they were probably right, and I was premature. I took off to do a Ph.D. in brain biochemistry and begin a research and teaching career, with a bit of politics on the side.

Two things challenged this complacency. One was on a global scale. The worldwide sweep of liberatory social movements that began in the late 1960s found me as a young post-doctoral neurochemist who suddenly saw, as the physicists had done before, that the findings of what I felt to be *my* science was being pressed into military service, this time in the genocidal American war in Indochina. History was making it clear before my eyes that science and politics were not separate, but intimately interwoven. The second change was personal. The early 1960s were the beginning of a relationship, of shared life and work, with the sociologist Hilary Rose. Science, she had argued, was too important a social phenomenon to be left to the scientists. Both its social origins and consequences needed to be analysed rather than taken for granted. As I began to learn this from her, I could also start to see the need to bring the separate fragments of my life together. Seeing the need and achieving the goal are two separate things though, and what sometimes has seemed to me to be the beginning of a synthesis has in retrospect too often felt inadequate or at best partial. Nonetheless, it is a goal which has determined the character of much that I have written since, whether alone or jointly with Hilary Rose, and it is the central message of all the essays collected in this book. For when I recently began to take stock of what I had been trying to say in a variety of ways to a variety of audiences over the last five years, I became very much aware that there was a pattern to it, in which I have been trying on the one hand to go beyond exploring the politics of science and the particular ideological role that scientific knowledge — especially biological knowledge — has come to play in the past two decades and on the other offering a more-or-less sophisticated account of the experimental work that has kept me in the lab on-and-off since the 1960s. Rather, I have been trying to explore what a new science might look like in the context of my own laboratory research.

In the first book we wrote together, *Science and Society*, Hilary Rose and I had tried to understand the nature of science as an institution in present day industrial societies like Britain by exploring its historical and social origins. By the seventies however, and with the experience of several years of debate within what had begun to be called the radical science movement, we had begun to argue, as others had done before and have done since, that it was not possible to understand the nature and power of science in the abstract, but instead it was necessary to regard it — the knowledge that it offers and the technologies it fosters — as a particular product of the society in which it is done; not neutral or outside the social order, but simultaneously a product and producer of social forms.

Within biology in particular, it seemed that the dominant mode of explanation sought was a reductionist one; of societies in terms of organisms, organisms in terms of cells, cells in terms of molecules and ultimately everything in terms of atoms. This reductionism serves a particular ideological role, because it goes hand in hand with a biological determinism which maintains that the social order with all its manifest inequalities is biologically ordained in accordance with the needs and instructions of our 'selfish genes'. During the 1970s and 1980s my writing has been concerned more and more with criticising this biological determinism. I have tried to show both its inadequacy as a way of accounting for the complexity of living systems and its ideological role in sustaining the *status quo*. (Scientific colleagues sometimes felt uneasy at this duality, arguing that it was enough to demonstrate, for instance, that sociobiology or IQ theory were poor biology and psychology without seeking to explore the social forces which made them significant; for my part I felt, and feel, that both are necessary tasks.)

But if one criticises reductionist biology when others do it, what should one be doing in the lab oneself? If, like me, you are interested in the seemingly almost intractable problem of exploring the cell biology of information processing in the brain, the biochemistry of learning and memory formation, you are brought abruptly in your own lab practice with attempting to cross the interface between biology and behaviour, between brain and mind. If you believe in the unity of the material universe, and that therefore there must be regularities in the relationship between descriptions of mind events and of brain events, how can you explore this interface without collapsing into a naive philosophical reductionism? The laboratory practice of the biochemist involves analysing the chemistry, metabolism and structure of brain cells, observing, for instance, how these change in animals when they are trained on novel tasks and learn new skills. Can such a project, so steeped in the *methodology* of reductionism, nonetheless avoid the philosophical and ideological baggage which reductionism brings in its train?

It is this problem to which I find I am returning again and again in my writing, and it is this which forms the central theme of the present set of essays. I want to describe how they came to be written and why I have assembled them in the present context, since most of them, although not all, have been published already in one form or another, sometimes in learned journals, sometimes in the context of conferences and meetings. However, I have reworked all of them to diminish the technical scientific content, to avoid overlap, and hopefully to aid coherence.

The book begins, as did my own thoughts in this area, with a discussion of the limits to science. Is research unbounded, and the researcher free to go where fancy takes? Is all knowledge eventually up for grabs, or are we constrained, by money, by goals, by ideology, by our own pasts? The origins of Chapter 1 lie in a debate held, under the title of 'The Limits to Science' between myself and James Watson at the Institute of Contemporary Arts in London. Watson himself, (perhaps unsurprisingly), took the position that science was, and indeed should be, unbounded except by the capacity of the scientist to design experiments, and argued an unabashed reductionism. In the last analysis, he claimed, 'What else is there but atoms?' By contrast, I argued — but why rehearse it here when it can be read in full in a few pages time? I have expanded the original text also to simplify the discussion of whether Western science can be regarded as independent of (neutral to) questions of race and gender, based on meetings I have been having with various groups of London science teachers who have been trying to create an anti-racist science syllabus.

The second and third chapters contain the most sustained discussion of the roots and ideological significance of reductionism and biological determinism in the present book. Both are extensions of themes raised in a book written jointly with Richard Lewontin and Leo Kamin and published by Penguin in 1984, *Not In Our Genes*. In Chapter 2 I begin by looking at types of explanation in biology, a theme that underpins many of the later chapters, and go on to discuss the historical origins of reductionism. I then set out in summary form what I see as reductionism's flaws as a totalising way of viewing the world. Chapter 3 is particularly concerned with the way in which biologically determinist ideas, especially those of sociobiology, have come to play a central part in the recent political thought and ideology of two different strands of New Right thinking, libertarian and authoritarian, not only in Britain but also in the USA and Europe. It is an extended version of an essay written jointly with Hilary Rose and first published in *Race and Class*.

Chapters 4 and 5 are both concerned, in different ways, with DNA and the new genetics. The issues are at once deeply theoretical and of profound practical import. The unravelling of the genetic code has given new insights — sometimes flawed insights — into the meaning and relationships of the concepts of genotype, phenotype and evolution in biology. The techniques of recombinant DNA (genetic engineering) have provided tools for the analysis of the human genome and apparently opened the path to gene therapy, to ways of 'perfecting' imperfect people. But what forms could such 'perfecting' take, and is the idea of perfectibility meaningful, let alone technically feasible? Meanwhile, what *is* not merely feasible but with us already, is an increasing military interest in the new genetic engineering techniques for the possible production of novel chemical and biological weapons systems, and Chapter 5 looks at the implications of this new arms race, at biotechnology at war. An earlier version of Chapter 4 was published in the *Lancet*, and of Chapter 5 in the *New Scientist*.

Chapter 6 looks at what has become one of the most emotive of issues surrounding contemporary biological research, the ethics of animal experimentation. Do animals have 'rights' and is it permissible to use them for experimental purposes? Some philosophers have become increasingly emphatic on this question in recent years, speaking to the concerns of a significant political movement. I try to apply the same criteria to the analysis of this issue as I have done to the general social uses and control of science, and to explain why I do not believe the concept of animal rights is helpful or even meaningful, and why I do not see any alternative to continuing with the use of some animals for some types of experimental and research purposes. This chapter has not been published in any form before.

In chapter 7 I turn to the theme which occupies me for the remainder of the book, that of transcending the paradoxes of mind/brain dichotomies in the creation of a unified but non-reductionist science of psycho- (or neuro-) biology. In 1980 a group of us calling ourselves the 'dialectics of biology group', were involved in a memorable conference that tried to pull together the strands of thought and practice that might be needed in the development of a non-reductionist science of the bio-social. The conference proceedings were published in a couple of volumes in 1982 (called *Against Biological Determinism* and *Towards a Liberatory Biology*). We always intended to have, but never succeeded in organising, a follow-up meeting. Indeed, it still seems like a good idea ... In any event Chapter 7 is a version of the paper I wrote for that conference. Beginning with the question of the meaning of *causation* in biology (an issue first addressed in Chapter 2) it introduces what for me is a key concept — that relationships between levels of description of phenomena, for instance, between brain and mind, cannot be understood as causal or merely correlative, but are best described as *translation* relationships between *correspondents* — that is, matching descriptions of the identical phenomenon at different levels. What this means, and how I believe it is an important part of transcending reductionism, I leave for the chapter to explain.

Chapter 8 describes what happens if translation relationships are mistakenly thought to be causal, in the sorry history of biological psychiatry's attempts to identify 'markers' of psychic disorder — that is, reliable biochemical indices of some metabolic disturbance which might underlie the disorder. I show that the blind alleys into which this research has led are not just the products of sloppy research but of poorly formulated questions which symbolise weak and mistaken theory about the nature of psychic distress and its relationship to biological processes. I was originally led to the survey of research which forms the background to this chapter by an invitation from the World Health Organisation to give one of the opening papers at a conference on biological markers in mental disorder held in 1983. This was a new venture for me. As a 'basic' neurobiologist I had up till then maintained a nodding acquaintance with the clinical literature, but had rapidly retreated to the relatively simpler problems provided by my chicks. Doing the WHO paper, however, forced me to think more seriously about the question of whether *any* rational biochemical approach to psychic distress was possible. The result was a new research approach that I am involved with at present, exploring the biochemical consequences of psychotherapeutic interventions. It is too early yet to say whether such research will be productive or illuminating. I may know more by the time this book is published.

Finally, then, to the main body of my own laboratory research on the cellular biology and biochemistry of learning and memory. In the last two chapters, I try to explain why I have chosen to spend the best part of the last twenty years working on this theme. It is because I believe that the concepts of learning and memory are central to our capacities as humans and our understanding of ourselves as individuals; and because I am convinced that an understanding of the cellular mechanisms of these fundamental brain/mind activities may prove to be the Rosetta stone which will allow us to decipher the translation relationships between the two. Chapter 9 (rewritten from a paper which originally appeared in the journal *Neuroscience*) discusses the general nature of the problem, and sets out what I believe to be the ground rules for a rational and appropriate research strategy in this area which, although reductionist in methodology, I claim is set within a non-reductionist theoretical framework. Chapter 10 (based on a review written for *New Scientist*) takes these ground rules, and tries to show how within my present lab practice I can actually apply them — what happens when I do, and what I now believe I know about how the brain makes memories.

Acknowledgements

The publishing history of the original texts which the chapters in this book have been written is as follows:-

Chapter 1 Originally from a debate at the ICA. An earlier version forms Chapter 3 of *Science and Beyond*, S. Rose and L. Appignanesi (eds), Blackwell, 1986.

Chapter 2 Derived from a chapter with the same title published as Chapter 1 of *More than the Parts*, L. Birke and J. Silvertown (eds), Pluto, 1984.

Chapter 3 Extended and edited version of an article of the same title, written jointly with Hilary Rose and published in *Race and Class*, *27*, 1986, pp. 42–64.

Chapter 4 An edited version of this chapter appeared, under a similar title,

in *Monthly Review*, *38*, 1986, pp. 48–60. It was itself derived from a more technical version of the paper in *The Lancet*, 15 December, 1984.
Chapter 5 An edited version of this chapter was published as Biotechnology at War, *New Scientist*, 19 March 1987.
Chapter 6 Unpublished hitherto.
Chapter 7 Derived from an article of the same name appearing as Chapter 1 of *Towards a Liberatory Biology*, S. Rose (ed), Allison and Busby, 1982.
Chapter 8 Extensively rewritten from a technical paper of the same title published in *the Journal of Psychiatric Research*, *18*, 1984, pp. 351–60.
Chapter 9 Extensively rewritten from a technical paper of the same title published as 'What should a biochemistry of learning and memory be about?' *Neuroscience*, 6, 1981, pp. 811–21.
Chapter 10 An edited version of this review was published as 'Molecules and Memories' in *New Scientist*, 27 November 1986.

1 The Limits to Science

For the great ideological 'spokesmen' of science, from Francis Bacon to James Watson, science has always been without limits; it is about 'the effecting of all things possible'. Human curiosity, after all, is boundless. There seems to be an infinity of questions one can ask about nature. At the end of his long scientific career Isaac Newton said he felt as if he had merely stood at the edge of a vast sea, playing with the pebbles on the beach. What is more, because science is not merely about passive knowledge of nature but also about developing ways of changing nature, and of transforming the world through technology, these same apologists offer us a breathtaking vision of the prospect of a world, a nature, including human nature, made over in humanity's image to serve human needs. It is only when one looks a little more closely at these visions that one sees that a science which claims to speak for the universality of the human condition, and to seek disinterestedly to make over the world to the benefit of humanity, is in fact speaking for a very precise group. Its universality turns out to be a projection of the needs, curiosities and ways of appreciating the world — not of some classless, raceless, genderless humanity — but of a particular class, race and gender who have been the makers of science and the framers of its questions from Francis Bacon onwards.

The ideology is powerful, and in the second half of this century has been of endless fascination to politicians as well as scientists. Towards the end of the Second World War, in the USA, Vannevar Bush, whose life had been spent with what he called '*Pieces of the action*' of science, offered Presidents Roosevelt and Truman '*Science, the Endless Frontier*' as a vision of how the greatness and power of the USA could be indefinitely extended. In Britain the visionary Marxist tradition of J.D. Bernal inspired Harold Wilson in 1964 to speak of 'building socialism in the white heat of the scientific and technological revolution' and Soviet scientists and politicians to speak of the 'scientific and technical revolution' which has, rather than politics and class struggle, become the motor of the growth of Soviet society.

Such claims for the limitless nature of human curiosity and the techno-economism of the politicians have been attacked by the critical and anti-science movement of the last decades — elements of the peace movement, feminists, the ecology and animal liberation movements among them. They have called for a halt to the 'tampering with nature' of the nuclear industry and militarism; a halt to the possibility of knowledge by the endless dissection of animals into molecules and molecules into elementary particles; a halt to the restless experimentation implied by the very scientific method itself as a way of knowing the universe, as opposed to the contemplative knowledge offered by alternative philosophical system.

I am not an 'anti-scientist' in this sense, or indeed in any other sense. I do want to argue, however, that we cannot understand science or speak of its limits or boundlessness in the abstract. To speak of 'science for science's sake — as if, to paraphrase Samuel Butler on art, science had a 'sake' — is to mystify what science is and what scientists do. This mystification, still often on the lips of the ideologues of science, serves to justify specific interests and privileges. Instead we have to consider *this* science in *this* society. This means that one has to understand the historical features of the emergence of what is now regarded as a universal science.

The orthodox interpretation of this history is that it has been a triumph of Western endeavour over the past three centuries. This interpretation needs questioning. Consider the conventional view within which most of us in the West have been trained as scientists. It is a view neatly expressed in the words of an ILEA school science inspector in 1985, when asked to advise on whether there could be considered to be a racial dimension to science:

> It is a truism which needs no emphasis that pure science is entirely neutral and that the community of scientists transcends national/ethnic boundaries. Consequently, the teaching of science *per se* is ethnically and culturally neutral. Scientific facts, principles and laws apply in a uniform manner in all societies in all parts of the world. As a result, there are few, if any scientific concepts or terms to which anyone could reasonably take exception on ethnic grounds.[1]

Thus, science is neutral, and it is only the technology which is not neutral. We can take nuclear power, and we can turn it into power stations or we can turn it into nuclear weapons, but nuclear power itself is neutral.

It is precisely this comforting framework which has to be challenged if we are to understand the peculiar features of science in contemporary society. The subsequent chapters of this book map aspects of this terrain in some detail; here I offer merely a guide to the issues. The point is that if science were neutral in this way then the problems which we are faced with in society today are problems to which one can apply the scientific method. If the wrong answers come out then this is because not enough science is being applied. On this argument we need more science to resolve the problems of pollution, of disease in contemporary society, of ending the arms race. If only we could divert the work of our scientists from military research and direct it towards socially useful technology, the problems of science and society may well be resolved. But I shall argue that it is not that simple.

Modern science as an organised study of the natural world developed in seventeenth-century Europe. It involved a method of inquiry, the asking of objective questions about the natural world, testing by experiment, making hypotheses, making theories, retesting those hypotheses and theories. On this view, the history of science is the story of progress. Old laws give way to new laws. New discoveries are made the whole time, unifying theories begin to be developed, and as a result of these theories, technological powers are unleashed. We move mountains, develop vast machines, make entirely new substances never found in nature. And now in the last part of the twentieth-century, in the middle of the biological revolution, we are going to fabricate and refabricate life itself by means of genetic engineering and molecular biology.

If it is true that science *as we know it* was born in north-western Europe in the seventeenth-century, then the question of the nature of that science becomes urgent. Why was it born in that particular way in Europe at that particular time? After all, there are many civilisations with a history of culture and scientific investigation which had preceded seventeenth-century Europe and had reached a high state of scientific and mathematical understanding, the Chinese and the Indian among others. What was it that was going on in Europe at that particular time which resulted in the invention of modern science as a method of understanding and changing the world? If one explores that history, it turns out to be contemporaneous with another extraordinary phenomenon, the invention of capitalism. The birth of capitalism and the birth of science in the West thus went hand in hand. First, because capitalism provided the impetus for the development of certain sorts of science and technology, and second, because science and technology provided the impetus for the capitalist mode of production and mercantile expansion. Galileo's mathematics was concerned not merely with models of the earth going round the sun but developing the mathematics for navigation and the firing of projectiles from guns.

Take that great discovery by Captain Cook and his surgeon James Lind, that in order to prevent scurvy among British sailors at sea, they had to be fed with extracts of lime juice (which is now known to contain Vitamin C). The conventional history of biology takes this as the starting point for the proof that a lack of Vitamin C causes scurvy. But there is another type of question to be asked about Cook's observation, which gives another type of answer to the question of the cause of scurvy. What does Cook's observation tell us about the reasons for the press-ganging of sailors into ships, long voyages and the grabbing of colonial territory? The history of the biology and medicine of the eighteenth-, ninteenth-, and twentieth- centuries is tied up with, and limited by, the practical questions set by the economic and social needs of the developing capitalist and colonial societies of Europe.

So what does limit today's science? I argue that its limits are now, as always, provided by a combination of two major, only partially separable, factors. The first is material, the second ideological.

The material factor is of course that of resource. Science costs money. In the advanced industrial countries of Europe — East and West — and the

USA, it consumes anything from two to three per cent of GNP. From 1945 to the late 1960s, science was expanding at an enormous rate, an exponential growth doubling every 10 to 15 years or so. In the 1960s, a historian of science, Derek de Solla Price, pointed out that the doubling rate had been constant from about the seventeenth-century on.[2] It became fashionable to calculate that by the twenty-first-century, every man, woman, child and dog in the world would be a scientist and the mass of published research papers would exceed that of the earth. But, like population growth, scientific growth could not continue unchecked.

Something had to stop, and indeed it did. In most countries from the late 1960s on, the growth of science as a proportion of GNP slowed, halted or even, as in Britain, was reversed. Sheer limitations of resource were limiting the growth of science. This trend can be seen in the development of the physicists' accelerators. First, each country had its own. Then there was the West European CERN project at Geneva and matching machines in the USA and USSR. Now, even if Britain were to stay in CERN, which is at present doubtful, the costs of a new generation of machines make the 'world accelerator' the logical next step. And beyond this? Just how much resource is going to be devoted to whirling particles around at speeds closer and closer to that of light? Boundless human curiosity is going to be bounded. The endless frontier closes.

A closer look at the funding of particle accelerators reveals another phenomenon. Ask high energy physicists why anyone should spend hundreds or thousands of millions of pounds on them, and you are likely to get the answer that it is high culture and society can afford it, like subsidising the opera. But they will be fooling themselves — or you. Because that is not the story they have been telling the politicians, who have gone on shelling out vast sums of money for physics since Hiroshima and Nagasaki in the not unreasonable belief that they would get bigger and better bombs or new sources of power out of the investment.

This brings us to the more important point about the material limits on science. For funding is not merely limited: it is *directed*. Of the 2–3 per cent of GNP Britain has spent on science since the 1950s, getting on 50 per cent, year in, year out, has gone on military research. The figure is now about 53 per cent, the highest for many years; this is much more, incidentally, than is spent by any other Western country except the USA; compare France's 35 per cent, Germany's 12 per cent or Japan's less than 5 per cent. If you want to know why so much scientific endeavour is directed to military ends, you must ask political questions about how the decisions are made. But there can be no doubt that this concentration on directing research towards military needs, and towards the industrial priorities of production and profit, profoundly shapes the direction in which science goes. Apologists for the purity of science (though it is the purest of high energy physics that gave us the bomb) may argue that this is all technology; they claim that real science is unaffected by such directive processes. They are on shaky ground making this science/technology distinction. The distinguished American organic chemist, Louis Fieser, invented that nastiest of conventional weapons, napalm, experimenting on it in the playing fields of Harvard during the Second World War. He wrote about his discovery afterwards in a fascinating book called simply *The Scientific Method.*[3] The argument that 'pure' science is free and undirected cannot be sustained for a moment.

Take the triumphant progress of molecular biology these past decades. There have always been two broadly contrasting traditions in biology, a reductionist, or analytic and atomising one, and a holistic or more synthetic one. This latter tradition was strongly represented in the 1930s by such developmental and theoretical biologists as Needham, Woodger and Waddington. There was a proposal to set up a major institute of theoretical biology in Cambridge which would have brought the field together. But the funding agency was to be Rockefeller, and Rockefeller, under the guidance of Warren Weaver and the British Medical Research Council, decided that the future was to be chemical. They backed biochemistry and molecular biology instead. The double helix, and all that followed from it from 1953 onwards, was a direct result of that funding decision. Many people would argue it was a correct one and I might well agree. The fact is that it changed the direction of biology by a deliberate act of policy. Rockefeller's decision is thus comparable to those being made routinely by government and charitable funding agencies as they decide which are high priority areas, and which should not be supported. The research councils, such as the Medical Research Council (MRC), the Science and Engineering Research Council (SERC) and the rest, have their priorities. They are, as it happens, still mainly molecular, even though most of the problems which the MRC ostensibly exists to help solve are clearly not going to be resolved by more molecular biology. Despite the massive funding of research into the molecular biology of cancer over the past decades — especially since the launching of the Nixon 'war on cancer' in the 1970s — it is clear that the most exquisite molecular research has brought us no nearer controlling a disease many of whose precipitating causes are located in the chemical environment of a drug-using industrial society. However, the vast funds Nixon allocated *have* given us more and more molecular biology.

Let me move from the material to the ideological limits to science. It is not just that we get the science we pay for, but that at a deeper level, what science we do. What questions scientists consider worth asking at any time, indeed, the very way they frame the questions, are profoundly shaped by the historical and social context in which we frame our hypotheses and realise our experiments. This can be discussed at three different levels.

First, we can only ask questions we can begin to frame; the role of chromosomes in cell replication and genetic transmission was unaskable until there were microscopes powerful enough to see the chromosomes, as well as a genetic theory to be tested — the technology and the theory came together at the beginning of the present century.

Second, not all scientific facts are of equal value. There is an infinity (in the strict sense of the term) of questions one can ask about the material world — although which ones are relevant and important is historically contingent. For example, in 1956 Sanger published the complete amino acid sequence of a protein, the first time anyone had done it. It took him about 10 years. The protein was insulin, and he got the Nobel Prize for sequencing it. That it was insulin, rather than any of the other 100,000-odd human proteins, or the many millions of other naturally occurring proteins, was fortuitous. It happened to be a relatively small molecule and it was available pure and in bulk. Within a few years several other proteins were sequenced, each time to a great, but diminishing scientific fanfare. Today anyone can do it within a few weeks with an automated machine. But is anyone going to *want* to determine the structure of *all* naturally occurring proteins, or even all human ones? There is a law of diminishing returns, to all except stamp collectors, and sometimes, Ph.D. students. So a new fact — the sequence of another protein — is nothing like as interesting as the first protein facts were. There is a limit to how many such facts are wanted, and most protein sequencing projects are scarcely worth a research grant these days; although there is now a serious proposal to spend some three billion dollars to sequence the DNA of the entire human genome, for apparently little other reason than that 'it is there' — revealing a type of Everest-complex designed to confirm biology's arrival among the 'big sciences'.

The third point is at a much deeper level than either of the other two. It seems to me there is a fundamental limit to the capacity of science — framed within the dominant paradigm in which most of us work — to give meaningful, let alone satisfying, answers to the great questions of human concern today. The issue of reductionism runs like a thread through many of the subsequent chapters of this book, so I will do no more than sketch the issues here.

The mode of thinking which has characterised the period of the rise of science from the seventeenth-century onwards is a reductionist one (Chapter 2). That is, it believes not merely that understanding the world requires disassembling it into its component parts, but that these parts are in some way more fundamental than the wholes they compose. To understand societies, you study individuals, to understand individuals you study their organs; for the organs their cells; for the cells their molecules; for the molecules their atoms ... right down to the most 'fundamental' physical particles. Reductionism is committed to the claim that it is *the* scientific method, that ultimately the knowledge of the laws of motion of particles will enable us to understand the rise of capitalism, the nature of love or even the winner of the next Derby. It also claims that the parts are ontologically prior to the wholes they compose.

The fallacies of such reductionism should be apparent. We cannot understand the music a tape-recorder generates simply by analysing the chemical and magnetic properties of the tape or the nature of the recording and playing heads — although these are *parts* of any such explanation. Yet reductionism runs deep. For Richard Dawkins and sociobiology the well-springs of human motivation are to be interpreted by analysis of human DNA; James Watson has argued 'What else is there but atoms?'[4] Well, the answer is the organising relations *between* the atoms, which are strictly not deducible from the properties of the atoms themselves. After all, quantum physics cannot even deal with the interactions of more than two particles simultaneously, nor predict the properties of a molecule as simple as water from the properties of its constituents.

Think of a Martian coming to earth and being confronted with the parts of an internal combustion engine. What are they for? The parts do not make sense by themselves, not even when they are reassembled into a car, unless you also know that the car is part of a transportation system. Yet why do scientists of experimental ingenuity and reputation consistently claim that you can understand the transportation system from the parts of the car engine? The roots of that belief go back, I think, to the Newtonian and Cartesian project for science as it has developed from the seventeenth-century. In Chapter 2, I try to show how reductionism was a scientific philosophy customised for capital's needs, and has remained so since. The trouble is that, just as capitalism was once a progressive force but has now become profoundly oppressive of human liberation, so too with reductionism. Beginning as a way of acquiring new and real knowledge about the world, from the structure of molecules to the motions of the planets, it has become an obstacle to scientific progress.

So long as science — in the questions it asks, and the answers it accepts — is couched in reductionist and determinist terms, understanding of complex phenomena is frustrated. I believe that a reductionist science cannot advance knowledge of brain functions, or solve the riddle of the relationship between levels of description of phenomena such as the 'mind/brain problem', which Western science is almost incapable even of conceiving except in Cartesian-dualist or mechanical-materialist terms. Reductionism cannot cope with the open, richly interconnected systems of ecology, or with integrating its scientific understanding of the present frozen moment in time with the dynamic recognition that the present is part of a historical flux, be it of the development of the individual or of the evolution of the species (Chapters 4, 8–10).

Failing to approach the complexity of such systems, reductionism resorts to more or less vulgar simplifications which, in the prevailing social climate, become refracted into defences of the status quo in the form of biological determinism, which claims that the present social order, with all its inequalities in status, wealth and power between individuals, classes, genders and races, is 'given' inevitably by our genes. This limit to the scientific vision is compounded by the closed recruitment process into science as an institution which effectively ensures its preservation as the privilege of the Western white male (Chapter 3).

I want to conclude this chapter by referring to one limit to science I have not yet mentioned, the *ethical* one. Ethical issues in science have been repeatedly discussed in recent years. They take several forms. On the one hand, some claims have been made that certain types of knowledge are too dangerous for humanity in its present state, and therefore some types of experiment should not be made. For instance, nuclear power, or gene cloning are considered to present hazards which make it inappropriate to pursue them experimentally (Chapters 4 and 6); alternatively, research on the so-called 'genetic basis of intelligence' might reveal biological 'facts' which would be unpalatable. On the other hand, it has been argued that the conduct of certain types of experiment, those which cause pain to animals for instance (or for that matter to humans), contravene absolute moral principles and should not be performed. All these considerations may be regarded as limiting science.

From what I have already said it should be apparent that I have a complicated response to this rather abstract approach to ethics. In my opinion, the resource and ideological questions are paramount; most ethical questions eventually break down to ones about economics, priorities and ideologies (see the discussion of *in vitro* fertilisation in Chapter 4 for example). Personally, I would not do research funded by, or with obvious applications to, the military. I will try to persuade as many of my fellow scientists as possible to take a similar ethical and political decision. But in the last analysis, in a militarist society *anything* one does can be and potentially will be co-optable for military purposes. If we do not want war-oriented research, individual ethical decisions are not enough. We need the *political* decision not to finance war research (Chapter 5).

Similarly (Chapter 6) I accept the case made by the animal liberationists that it is undesirable to use procedures likely to cause pain or distress to animals, although in the last analysis I owe my prior loyalty to my own species, and to argue otherwise seems perverse. I care more about saving people than saving whales. But a vast proportion of the animal experimentation done in Britain is either for relatively trivial commercial purposes — developing new drugs for instance, when it is at least arguable that there are enough or even too many drugs available already; what is needed is not new magic drugs but a health-producing society. And indeed many drugs which are developed and tested are not new in a fundamental sense but are part of the endless process of molecular roulette played by the drug companies in their efforts to circumvent patent laws or maximise profits. It is also true that a fair number of the animal experiments done in 'basic science' labs are, on close analysis, carried out in the pursuit of trivial or 'me-too' type research aims. Remember that the average scientific paper is probably read by only one or two other people apart from the editor of the journal in which it appeared and the referees. So part of my answer to the question of ethics and animal experiments is to rephrase the question in terms of whether the research is worth doing anyhow, animals or no.

So too with the question of 'things we are not meant to know'. These are often just things it isn't worth trying to know, like the sequence of every

ALLEGHENY COLLEGE LIBRARY

possible naturally occurring protein. But sometimes they are things which *cannot* be known because the questions are simply wrongly or meaninglessly phrased. As someone who has been involved in what has become known as the 'race-IQ' debate I have often been asked whether I am opposed to work on 'the genetics of average race differences in IQ' on ethical grounds. My response is that I am opposed to it on the same grounds that I am opposed to research on whether the backside of the moon is made of gorgonzola or of stilton. That is, it is a silly question, incapable of scientific answer and actually, *sensu strictu*, meaningless. The question makes grammatical, but not scientific sense, because 'IQ' is not a phenotype susceptible to genetic measurements, and heritability estimates cannot be applied to average differences in phenotypes between groups.

All this is not to duck the questions of ethics. There are issues of real choice and dilemma in medicine, in the use of animals, and indeed in some aspects of biotechnology, which cannot simply be reduced to issues of economics and ideology. They are few but important, and they set limits to our science. How should they be resolved? In the last analysis, not by scientists playing god-in-white-coat and refusing to allow anyone else in on the decision. And not by committees of professional ethicists and philosophers. The only way of dealing with such issues is by democratic participation in decision-making about what science is done. My own aim would be for a way of controlling and directing research which opened up all laboratories to community involvement in their direction; and planned work by a combination of the tripartite structure of decision making by scientists and technicians in the lab itself, by the community in which the lab was embedded and by discussion of overall priorities and resources at a national level. I believe that if we did organise our science in this way, not merely would new priorities set different limits to our work, but that we might also begin to see the makings of a new, less reductionist and more holistic, human-centred science.

Notes and References

1. Statement made by an HMI faced with the challenge of designing an anti-racist science curriculum, quoted in ILEA Science Centres: *Science Teaching in a Multi-ethnic Society*, ILEA, 1986.
2. D. J. de Solla Price, *Little Science, Big Science*, Columbia UP, New York 1963.
3. L. Fieser *The Scientific Method*, Reinhold, New York, 1964.
4. These claims were made by Dawkins and by Watson in the series of debates at the ICA the first one of which forms the basis for the present chapter.

2 Biological Reductionism: Its Roots and Social Functions

Ask biologists what they are trying to do when they record data, make experiments or propose theories. There are obvious answers, like earning a living, solving a problem set by the research director or completing a Ph.D. Go beyond those and the answer is likely to be that they are trying to find out how or why a particular phenomenon occurs in an ecosystem, a living organism or a subcellular preparation. But what sort of answer is expected to questions of why? and how? What would be regarded as a satisfactory explanation of a particular biological event?

Biology is a strange amalgam of different disciplines. The organism that the biochemist or the physiologist or the geneticist studies may well be the same: the differences between these disciplines lie in the questions they ask of the organism and the level of analysis at which they study it. For the biochemist, the organism is a sort of sack of molecules to be extracted and purified, and their composition and interactions determined. To the physiologist, it is a mass of cells organised into discrete but interacting organs; for the ecologist, it is not the individual but the population of organisms that is important. The geneticist asks why one organism is different from another

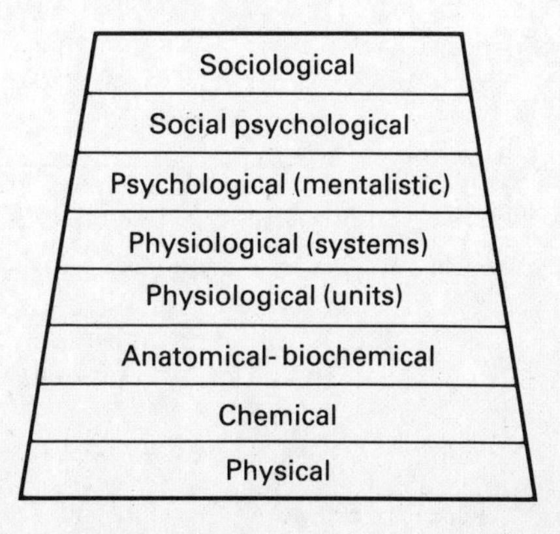

Fig. 1 The conventional hierarchy of science

of the same species; the developmental biologist wonders how the organism grows from a single cell or embryo into its characteristic adult form. What is the relationship between all these different questions and levels of analysis?

It is conventional to speak of levels of analysis in biology as if they were arranged in a hierarchical order, as in Fig. 1. Going downwards through the hierarchy there is a movement, so the convention claims, in the direction of increasingly fundamental components, towards the atomic and the sub-atomic; whereas in moving upwards, the worlds became increasingly complex up to the social domain of the interactions of individuals and populations. Given such a hierarchy, how can we set about explaining a phenomenon at any given level?

Take a simple example: the contraction of a muscle in a frog's leg. There are five types of possible answers that biologists might give to the question. 'What caused the contraction'? Their relationship is summed up in Fig. 2.

'Within-level' explanations

One might, for example, say that the frog's muscle twitched *because* an appropriate set of impulses passed down the motor nerve innervating it, which signalled the instruction to contract. This sort of explanation will describe a present phenomenon as *caused* by an immediately prior event. First the nerve fires, and then the muscle twitches — and we can go on to explain that the nerve fired as a result of some earlier appropriate set of inputs to its motor neurons, derived from the frog's brain and/or its sensory input. So we have a sequential series of events that follow one another in time and are linked in a transitive and irreversible way. That is, *first* event A occurs; *then* as a result, event B; and as a result, event C, and so on. This is a

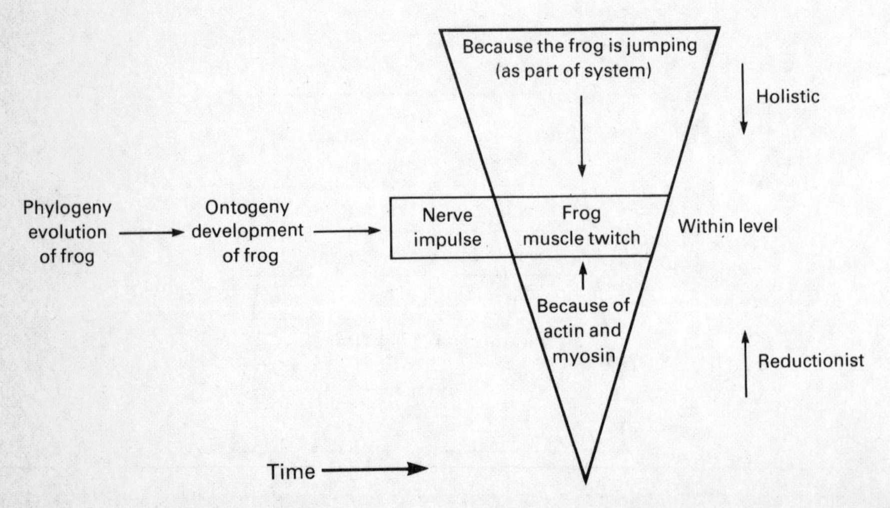

Fig. 2 The relationship between the types of biological explanation

straightforward causal chain, with all the individual components described in the same language and within a single level of analysis — that of the physiologist. The sequence can be summarised thus:

external events → sensory stimulation → brain integration → motor nerve firing → muscle twitch

The single headed arrows emphasise that the sequence cannot run backwards, that is, the muscle twitch cannot cause firing of the motor nerve.

'Top-down' explanations

One can, however, consider the activity of the whole organism, and then state 'the muscle twitched *because* the frog was jumping to escape a predator'. Here the explanation of the activity of part of a complex system is given in terms of the integrated functioning of the system as a whole. Such systems explanations are a way of describing the goal directedness of an organism (teleonomy) without implying the concept of purpose which makes many philosophers uneasy.

They give meaning to a phenomenon or structure which cannot be given any other way; to consider the heart without recognising it as part of a circulatory system would clearly prevent one from ever properly understanding it. But note a particular aspect of this type of explanation. 'The muscle twitched *because* the frog was jumping' is a different class of statement from 'the muscle twitched *because* a signal arrived from a motor nerve'. The latter statement as we saw above, is about causal chains linked in time: *first* the nerve fires and *then* the muscle contracts. But we do not say *first* the frog jumps and *then* the muscle twitches. The two events are not sequential; rather the use of the word 'cause' here implies a logical connection, not a temporal one. Normally frogs are unable to jump unless during this jump, their leg muscles twitch; twitching is part of the activity of jumping. So the types of top-down explanation we can offer are different in kind from those within a single level.

But there are problems in integrating top-down explanations, or *holistic* as they are generally known, [1] with other types of explanation, and there is much confusion about them. Some biologists have regarded them as quite improper, because of the confusion of the two senses in which the word 'cause' is used. Others claim that only holistic explanations can really be satisfactory, that there is a process of 'downward causation' by which the properties of the system — the organism — *constrain* or *determine* the behaviour of the parts. The system becomes thus 'more important' than the parts of which it is composed. If an experimenter severed the motor nerves to the frog's leg muscle, or paralysed the muscle with a chemical poison, the frog would still endeavour to escape its predator, possibly successfully, by employing a different set of muscles or a different escape strategy. To the goal-directed organism there are multiple paths to a given end.

'Bottom-up' explanations

Holism bears a sort of mirror-image relationship to the type of explanation which is the main topic of this chapter, that of *reductionism*. Consider the frog muscle. It is itself composed of individual muscle fibres. These themselves are largely composed of fibrous proteins. In particular, there are two types of protein, actin and myosin, arranged within the muscle fibril in characteristic arrays. When muscle fibrils contract, the actin and myosin chains interdigitate; the conformational change as they slide between one another involves the expenditure of energy, and a chemical substance, ATP, is broken down in the process. So a bottom-up explanation of the muscle twitch would be in terms of the proteins of which the muscle is composed. It contracts *because of* the protein filaments sliding past each other. It is possible to go on to explain the conformational changes in these proteins in terms of the amino acid composition of the individual actin and myosin molecules, and the molecular interactions of group transfer molecules like ATP. These in turn could be described in terms of the atomic or quantum structure of the molecules, and so on.

Like holism, then, reductionism is a *between-levels* explanation. It is the most commonly accepted type of explanation offered by biologists. The scientific culture of contemporary Western society, with its belief in the hierarchy of sciences of Fig. 1, teaches that reductionism offers 'more fundamental' explanations than any other. Higher order levels are, it is argued, in the long run (if not at present) to be explained in terms of the lower order levels.

Indeed, some reductionists go on to claim that *the* task of science is to dissolve the higher orders completely into the lower orders. This type of thinking is particularly strong among biochemists and molecular biologists like Jacques Monod and Francis Crick. Francis Crick has argued that all the important biological questions can be 'solved' by concentrating on unravelling the molecular architecture and dynamics of the bacterium *E.coli*. It was Crick who formulated what he described as the 'central dogma' of molecular biology, the unilinear chain of causation, which says that information flows only *from* the genetic material, from DNA to protein, and hence to the organism's phenotype, but not in the reverse direction, from phenotype to DNA.

Reductionism is a powerful explanatory principle in biology, and we are easily led to believe that such reductionist explanations are the most important, or indeed that they are the only 'real' scientific ones. Most of us find the idea that the movement of actin and myosin sliding past one another is the *cause* of the muscle contraction less difficult than the idea that the contraction was *caused* because the frog was attempting to escape from a predator. Many cell biologists would feel uneasy with the latter teleonomic form. Yet formally, reductionist 'causation' is of the same type as holistic 'causation' and both are unlike within-level causation. That is to say, there are not *two* sequential events, *first* the passage past one another of the actin and myosin chains and *then* the muscle contraction. Both happen simultaneously. The passage of the actin and myosin chains past one another is just another way of describing muscle contraction. There is a single event or phenomenon which can be described at one of two different levels. The muscle contraction is *identical* to the sliding of the actin and myosin chains. Thus the relationship between the two descriptions, of muscle contraction and of actin and myosin chains sliding past one another, is not *temporal*, as offered by the within-level explanations; rather it is one of *logical necessity* (I take this point up in more detail in Chapter 7). The muscle fibril is an ensemble of action and myosin molecules, and when it contracts the configuration of these molecules alters, by definition.

Developmental and evolutionary explanations

Before looking at the origins and social function of reductionism, it is necessary to refer briefly to the two other types of explanation of the frog muscle twitch shown in Fig. 2. Developmental explanations would attempt to explain the twitch of the muscle in terms of the sequence of events by which muscle cells become specialised from the early embryo and by which they become attached at the two ends and innervated by appropriate nerves, so that the specificity of nerve-muscle connections is achieved and maintained, and the twitch then produces certain defined movements. Such explanations are concerned with how the orchestration of various muscles and nerves is achieved so that their actions are co-ordinated into an appropriate output. A developmental explanation differs from a physiological one in that it tells an historical story; the causal sequence begins not merely with the firing of the motor nerce but before the birth of the organism. Hence the time-scale of the explanation offered is much longer and the purpose it must then serve in terms of accounting for a postulated activity in terms of the unrolling past of the organism is very different.

Evolutionary explanations are concerned also with temporal sequences of events and historical causations for a present situation, but compared with developmental explanations, the time sweep of such attempted explanations transcends the lifetime of any individual and looks at that of the species as a whole. Such explanations have been very popular in recent years, often in the guise of adaptationism or sociobiology, to which I return below.

The origins of reductionism

Having surveyed the five types of explanations offered by contemporary biology, I now return to the question of the origins of reductionist thought.

My thesis is that it arose with the birth of modern science in seventeenth-century Europe, and that its history is intimately connected with the development of a particular world view. The rise of modern physics, first with Galileo and then particularly with Newton, ordered and atomised the natural world. Beneath the surface world in all its infinite variety of colours, textures, and varied and transient objects, the new science found another world of absolute masses interacting with one another according to invariant laws which were as regular as clockwork. Causal relationships linked falling bodies, the motion of projectiles, the tides, the moon, and the stars. Gods and spirits were abolished or relegated merely to the 'final cause' which set the whole clockwork machinery in motion. (Actually Newton himself remained both religious and mystic throughout his life, but that is one of the minor quirks of personal history: the effect of Newtonian thought was the reverse of Newton's personal philosophy.) By contrast with the feudal world, the universe thus became de-mystified and, in a manner, disenchanted as well.

The post-Newtonian world that emerged was one in which once again heavenly and earthly orders were in seeming harmony. The new physics was dynamic and not static, as were the new processes of trade and exchange which came with the development of the new capitalist economic system. There was a set of new abstractions to describe the world in which a series of abstract forces between atomistic and unchanging masses underlay all transactions between bodies. Drop a pound of lead and a pound of feathers from the leaning tower of Pisa, and the lead will arrive at the ground first because the feathers will be more retarded by air pressure, frictional forces, and so on. But in Galileo's and Newton's equations, the pound of feathers and the pound of lead arrive simultaneously because the *abstract* pound of lead and pound of feathers are equivalent unchanging masses to be inserted into the theoretical equations of the laws of motion.

These abstractions paralleled the world of commodity exchange in which the new capitalism dealt. To each object there are attached properties of mass or value, which are equivalent to or can be exchanged for objects of identical mass or value. Commodity exchange is timeless, unmodified by the frictions of the real world; for example, a coin does not change its value by passing from one hand to another, even if it is slightly damaged or worn in the process. Rather, it is an abstract token of a particular exchange value. It was not until the nineteenth-century that this view could become fully dominant. Joule demonstrated that all forms of energy and heat, electromagnetism and chemical reactions were interchangeable and related by a simple constant, the mechanical equivalent of heat; and later Einstein demonstrated the equivalence of matter and energy. These demonstrations corresponded to an economic reductionism whereby all human activities could be assessed in terms of their equivalents in pounds, shillings and pence. Nature and humanity itself had become a source of raw materials to be extracted by applying science and technology — which expertise remained firmly in the hands of the dominant class and gender. The transition from the pre-capitalist world of nature could not be more complete.

So far I have discussed physics as though it were all of science. But where did the new mechanical and clockwork vision of the physicists leave the status of living organisms? Just as modern physics starts with Newton, so modern biology must begin with Descartes, philosopher, mathematician and biological theorist.

For Descartes the world was machine-like, and living organisms merely particular types of clockwork or hydraulic machines. It is this Cartesian machine image which has come to dominate science and to act as the fundamental metaphor legitimating the bourgeois world view of a mechanical nature. That the machine was taken as a model for the living organism — and not the reverse — is of critical importance. Bodies are indissoluble wholes that lose their essential characteristics when they are taken to pieces. Machines, on the contrary, can be disarticulated to be understood and then put back together. Each part serves a separate and analysable function, and the whole operates in a regular, law-like manner that can be described by the operation of its separate parts impinging on each other.

Descartes' machine model was soon extended from non-human to human organisms. It was clear that most human functions were analogous to those of other animals and, therefore, were also reducible to mechanics. However, humans had consciousness, self-consciousness and a mind, which for Descartes, a Catholic, was a soul — and by definition the soul, touched by the breath of god, could not be a mere mechanism. So there had to be two sorts of stuff in nature: matter, subject to the mechanical laws of physics, and soul or mind, a non-material stuff which was the consciousness of the human individual, their immortal fragment. Descartes speculated that mind and matter interacted by way of a particular region of the brain, the pineal gland, in which the mind/soul resided when incorporate, and from which it could turn the knobs, wind the keys, and activate the pumps of the body mechanism.

So developed the inevitable but fatal disjunction of Western scientific thought, the dogma known in Descartes' case and that of his successors as 'dualism'; a dogma which, as we shall see, is the inevitable consequence of any sort of reductionist materialism which does not in the end wish to accept that humans are 'nothing but' the motion of their molecules. Dualism was a solution to the paradox of mechanism which would enable religion and reductionist science to stave off for another two centuries their inevitable major contest for ideological supremacy. It was a solution which was compatible with the capitalist order of the day because in weekday affairs it enabled humans to be treated as mere physical mechanisms, objectified and capable of exploitation without contradiction, while on Sundays ideological control could be reinforced by the assertion of the immortality and free will of an unconstrained incorporeal spirit unaffected by the traumas of the workday world to which its body had been subjected.

The development of a materialist biology

For the confident and developing science of the eighteenth- and nineteenth-centuries, dualism was but a stepping stone towards a more thoroughgoing mechanical materialism. The demonstration by Lavoisier that the processes of respiration and the sources of living energy from the oxidation of foodstuffs in the body tissues, were exactly analogous to those of the burning of a coal fire, was perhaps the most striking vindication of this approach. It was the first time that the programmatic statement that life must be reducible to molecules could be carried into practice.

But progress in the identification of body chemicals was slow. The demonstration that the substances of which living organisms are composed are only 'ordinary' albeit complicated chemicals came early in the nineteenth-century. The intractability of the giant biological molecules — proteins, lipids, nucleic acids — to the analytical tools then available remained a stumbling block. The mechanists could make programmatic statements about the reducibility of life to chemistry, but these remained largely acts of faith. It was not until a century after the first non-organic synthesis of simple body chemicals that the molecular nature and structures of the giant molecules began to be resolved (and really not until the 1950s that progress became very rapid). The last remaining notion that there would be some special 'life-force' operating among them which distinguished them absolutely from lesser, non-living chemicals lingered until the 1920s.

Nonetheless, a radically reductionist programme characterised the statements of many of the leading physiologists and biological chemists of the nineteenth-century. All bodily processes were to be described in physico-chemical terms. Moleschott and Vogt as thoroughgoing mechanical materialists, claimed that humans are what they eat, genius is a question of phosphorus and that the brain secretes thought as the kidney secretes urine. Virchow, one of the leading figures in the development of cell theory, was also part of a long tradition of social thought which argued that social processes could be described by analogy with the workings of the human body.

It is important to understand the revolutionary intentions of this group. They saw their philosophical commitment to mechanism as a weapon in the struggle against orthodox religion and superstition. Several of them were also militant atheists, social reformers, or even socialists. Science, they believed, would alleviate the misery of the poor and strengthen the power of the state against the capitalists, and even, in some measure, help democratise society. Their claims were part of the great battle between science and religion in the nineteenth-century for supremacy as *the* dominant ideology of bourgeois society, a fight whose outcome was inevitable but whose final battlefield was to be that of Darwinian natural selection rather than physiological reductionism. The best known philosopher of the group was Feuerbach, and it was against his version of mechanical materialism that Marx launched his famous theses.

The theses on Feuerbach proved the starting point for Marx's own, and more explicitly Engels' long-running attempts to transcend mechanical materialism by formulating the principles of a materialist but non-reductionist account of the world and humanity's place within it: dialectical materialism. But within the dominant perspective of biology in the Western tradition, Moleschott's mechanical materialism was to win out, stripped of its millenarian goals and, by the late twentieth-century, revealed as an ideology of domination. When biochemists today claim that a disordered molecule produces a diseased mind', or psychologists that inner-city violence can be cured by cutting out sections of the brains of ghetto militants, they are speaking in precisely this Moleschottian tradition.

To complete the mechanical materialist world picture, however, a crucial further step was required involving the question of the nature and origin of life itself. The mystery of the relationship of living to non-living presented a paradox to the early mechanists. If living beings were 'merely' chemicals, it should be possible to recreate life from an appropriate physico-chemical mix. Yet one of the biological triumphs of the century was the rigorous demonstration by Pasteur that life only emerged from life; spontaneous generation did not occur. The resolution of this apparent paradox awaited the Darwinian synthesis, which was able to show that although life came from other living organisms and could not arise spontaneously, each generation of living things might change and evolve as a result of the process of natural selection.

With the theory of evolution came a crucial new dimension to the understanding of living processes, the dimension of time. Species were not fixed immemorially but were derived in past history from earlier forms. Trace life back to its evolutionary origins and one could imagine a primordial warm chemical soup in which the crucial chemical reactions could occur. Living forms could coalesce from this pre-biotic mix. Darwin speculated about such origins, although the crucial theoretical advances depended on the biochemist Oparin and the biochemical geneticist Haldane in the 1920s, both, incidentally, consciously attempting to work within a dialectical and non-mechanist framework. Experiments only began to catch up with theory from the 1950s onward.

The consequence of Darwin's theory of evolution was to finally change the form of the legitimating ideology of bourgeois society. No longer able to rely upon the myth of a deity who had made all things bright and beautiful and assigned each to their estate, the rich ruler's castle or the poor peasant's gate, the dominant class de-throned God and replaced him with science. The social order was still to be seen as fixed by forces outside humanity, but now these forces were natural rather than deistic. If anything, this new legitimator of the social order was more formidable than the one it replaced. It has, of course, been with us ever since.

Natural selection theory and physiological reduction were explosive and powerful enough statements of a research programme to occasion the replacement of one ideology, religion, by another — a mechanical, materialist science. They were, however, at best only programmatic, pointing along a route which they could not yet trace. For example, in the absence of a theory of the gene, Darwinism could not explain the maintenance of inherited variation that was essential for its operation. The solution awaited the development of modern genetical theory which took place at the turn of the twentieth-century. This in turn produced the neo-Darwinian synthesis of the 1930s and the recurrent attempts to parcel out biological phenomena into discrete and essentially additive causes, genetic and environmental — the science of biometry.

The central dogma: centrepiece of the mechanist programme

Mechanist physiology took even longer to triumph. It depended on the development of powerful new machines and techniques for the determination of the structure of the giant molecules, for observing the microscopic internal structure of cells, and, above all, for studying the dynamic interplay of individual molecules within the cell. By the 1950s it had begun to be possible to describe and account for (in the mechanistic sense) the behaviour of individual body organs — muscle, liver, kidneys — in terms of the properties and interchange of individual molecules — the mechanist's dream.

The grand unification between the concerns of the geneticists and those of the mechanist physiologists came in the 1950s, the 'crowning triumph' of twentieth-century biology: the elucidation of the genetic code. This required a theoretical addition to the mechanistic programme, to be sure. Hitherto it had been sufficient to claim that a full accounting for the biological universe and the human condition was possible by an understanding of the trio of *composition*, the molecules which the organism contains; *structure*, the ways in which these molecules are arranged in space; and *dynamics*, the chemical interchanges among the molecules. To this now needed to be added a fourth concept, that of *information*.

The concept of information itself has an interesting history, arising as it did from attempts during the Second World War to devise guided missile systems, and through the 1950s and 1960s in laying the theoretical infrastructure for the computer and electronics industries. The understanding that one could view systems and their actions in terms not merely of matter and the energy flow through them but in terms of information exchanges, that molecular structures could convey instructions or information one to the other, shook up a theoretical kaleidoscope. In one sense this made possible Crick and Watson's recognition that the double helical structure of the DNA molecule could also carry genetic instructions across the generations. Molecules, the energetic interchanges between them *and* the information they carried, provided the mechanist's ultimate triumph, expressed as already mentioned in Crick's deliberate formulation of what he called the 'central dogma' of the new molecular biology, that there is a one-way flow of information between DNA and protein, a flow which gives historical and ontological primacy to the hereditary molecule. It is this which underlies the sociobiologists' 'selfish gene' argument that, after all, the organism is merely DNA's way of making another DNA molecule; that everything, in a preformationist sense that runs like a chain through several centuries of reductionism, is in the gene.

It is hard to overemphasise the ideological organising function fulfilled by this type of formulation of the mechanics of the transcription of DNA into protein. The imagery of the biochemistry of the cell, long before Crick, had been that of the factory, a factory whose functions were specialised for the conversion of energy into particular products and with its own part to play in the economy of the organism as a whole. Some ten years earlier than Crick's formulation, Fritz Lipmann, discoverer of one of the key molecules engaged in energy exchange within the body, ATP, formulated his central metaphor in almost pre-Keynesian economic terms: ATP was the body's energy currency. Produced in particular cellular regions, it was placed in an 'energy bank' in which it was maintained in two forms, those of 'current account' and 'deposit account'. Ultimately, the cell's and the body's energy books must balance by an appropriate mix of monetary and fiscal policies.

Crick's new metaphor was more appropriate to the sophisticated economics of the 1960s in which considerations of production were diminishing relative to those of its control and management. It was to this new world that information theory with its control cycles, feedback and feedforward loops, and regulatory mechanism was so appropriate, and it is in this new way that the molecular biologists conceive of the cell — an assembly-line factory in which the DNA blueprints are interpreted and raw materials fabricated to produce the protein end-products in response to a series of regulated requirements. Read any introductory textbook to the new molecular biology and you will find these metaphors as a central part of the description of cells. Even the drawings of the protein synthesis sequence itself are often deliberately laid out in 'assembly-line' style. The metaphor not only dominates the teaching of the new biology: it and language derived from it are key features of the way molecular biologists themselves conceive of and describe their own experimental programmes.

And not merely molecular biologists. The synthesis of physiology and genetics provided by an information theory containing a double helix was steadily extended upward from individuals to populations and their origins by biologically determinist sociobiology, which draws explicitly on molecular biology's central dogma to legitimate the claim that the gene is ontologically prior to the individual, and the individual to society.

For the mechanical materialists the grand programme which was begun by Descartes has now in its broad outline been completed; all that remains is the filling in of details. Even for the workings of so complex a system as the human brain and consciousness, the end is claimed to be in sight. An immense amount is known about the chemical composition and cellular structure of the brain, about the electrical properties of its individual units, and indeed of great masses of brain tissue functioning in harmony. Neurobiologists claim to have shown how the analyser cells of the visual system, or the withdrawal reflex of a slug given an electric shock, are wired up, and to have found regions of the brain whose function is concerned with anger, fear, hunger, sexual appetite or sleep. The mechanist's claims here are clear. In the nineteenth-century, Darwin's supporter, T.H. Huxley, dismissed the mind as no more than the whistle of the steam train, an irrelevant spin-off of physiological function. Pavlov, in discovering the conditioned reflex, believed he had the key to the reduction of psychology to physiology, and one brand of reductionism has followed him. For this tradition, molecules and cellular activity cause behaviour, and as genes cause molecules, the chain which runs from particular unusual genes, say, to criminal violence and schizophrenia, is unbroken.

So what's wrong with reductionism?

I have tried to show how reductionism emerged in intimate relationship with the rise of capitalism. But in doing so I have implied that this is an inadequate, painfully blinkered way of viewing the world, despite the grandiose claims for universal synthesis offered by its ideologues.

Why won't reductionism work? I want to argue that there are a series of major flaws in reductionist thought. It is sometimes claimed that opponents of reductionism object to it only on political grounds, or as applied to humans but not to other animals – as if Descartes was alive and well among us. While the political positions derived from reductionism, so in accord with the ideology of the New Right in politics – from Thatcher in England and Reagan in the USA to la Nouvelle Droit in France and the writings of Italian fascism – are indeed the enemy, reductionist philosophies need to be confronted on their own terrain, irrespective of the politics they succour or espouse. If reductionism as a method of explaining the world is flawed, it is flawed for other animals as well as humans.

The sociobiologist Richard Dawkins in his latest book, *The Blind Watchmaker* has offered as a defence of reductionism, a distinction between what he calls 'stepwise' and 'precipice' reductionism. Stepwise reductionism merely seeks 'causal' explanations for phenomena at a level one below that being studied – for example, muscle contraction in terms of the interaction of actin and myosin filaments. By contrast precipice reductionism goes the whole hog, explaining the frog jumping in terms of the genes which code for actin and myosin proteins. The first form is, he argues, what good science is all about. The second form is not seriously believed by anyone, he claims. However, this cheerfully slapdash pragmatism really won't bear examination. Logically, stepwise reductionism implies precipice reductionism as one proceeds 'downwards' through the levels of organisation of biological systems (I come back to this in Chapter 7). And Dawkins himself regularly and unabashedly employs the language of precipice reductionism throughout the book. Consider for instance his claim that

> It is raining DNA outside ... at the bottom of my garden is a large willow tree, and it is pumping downy seeds into the air.... mostly made of cellulose, and it dwarfs the tiny capsule that contains the DNA, the genetic information ... (but) ... it is the DNA that matters.... The whole performance, cotton wool, catkins, tree and all, is in aid of one thing and one thing only, the spreading of DNA around the countryside. Not just any DNA, but the DNA whose coded characters spell out specific instructions for building willow trees ... It is raining instructions out there; it's raining programmes; it's raining tree-growing, fluff-spreading algorithms. That is not a metaphor, it is the plain truth. It couldn't be any plainer if it were raining floppy discs.[2]

I don't see how anyone *but* a precipice reductionist could write so flagrantly.

I argue that the flaws in reductionism as an epistemology are not trivial, or confined to precipices, but fundamental (I am not here criticising *methodological* reductionism, a way of doing experiments in which one endeavours to control all but one or two variables). As a philosophy, reductionism begins by an assertion of ontological priority. That is, it claims that the individual is ontologically prior to the society of which that individual is a member, and the atom is ontologically prior to the organism. To go back to the example of the frog muscle twitching, it insists either that *first* actin and myosin molecules interdigitate and *then* the muscle twitches, or that in some sense the explanation of the twitch in terms of molecules is more fundamental than any other explanation, the only 'true' explanation, the goal of 'real science'. Reductionism thus cannot accept that phenomena are *simultaneously* both individual and part of a greater unity. Reductionism in the sociobiological sense begins from this philosophical premise of the ontological priority of the individual over society. In the biochemical sense, it insists on the priority of the molecule over the organism.

Secondly, reductionism operates by a phenomenon of arbitrary agglomeration and reification. Consider the way in which its protagonists use concepts like *aggression*, in which they lump together social phenomena such as war, strikes, discord between the sexes, football hooliganism, the space race, or whatever, as if they were all expressions of the one unitary phenomenon, aggression. What is more, the phenomenon labelled as 'aggression' in society is simply seen as the sum of the aggressive properties of the individual members of that society. One sees how this works in the context of the various medical interventive strategies offered to cure 'aggression' in society, notably the surgical interventions proposed after the inner-city riots in the USA from the 1960s onwards. Such strategies claimed that the explanation for such social phenomena must lie within the individuals involved in the aggressive activity. Therefore one should intervene, and 'cure' the riots by performing appropriate surgical operations on the brains of selected ghetto ringleaders to remove the 'aggressive centres' within them.

The process then has been first to assert the ontological priority of the individual over the society and then arbitrarily agglomerise different phenomena under a single label. The third step is to say that if this 'aggression' (or whatever) is a property of the individual, it must be located 'somewhere' in the body. There is a powerful homunculus model which operates through reductionist thinking, so that properties of the organism have to have a localisation in the brain; there must be a *site* in the brain for intelligence, a *site* in the brain for aggression, for sexuality, and so on. Yet point localisation of individual behaviour within a set of cells or molecules within the brain, is a fallacious way of understanding how brains — or individuals — actually operate. Behaviour is an expression of the properties of the system, of the organism; it is not located in any one part thereof. Consider a lecturer speaking into a microphone, which gives, instead of amplified speech, a howl. Where is the howl located? Not in the amplifier, or the microphone on loudspeakers, but in a reasonating feedback loop between them all.

The next step that reductionism takes (and it is a very powerful trend within the history of Western science) is a process that can be referred to as arbitrary quantification. This is the belief that it is possible to quantify any particular property, whether it is aggression, or intelligence, or whatever, and rank it upon some linear scale, so that one can state of any individual that he or she is x per cent more intelligent, or more aggressive, than another individual. This process of quantification expresses a naive belief in the power of algebra and physics to explain phenomena. There is an assumption that all such properties are linear, and can be mapped on some sort of scale. Again, let us take aggression as an example. Here is an example of how 'aggression' has now been studied in humans by way of the analysis of aggression in animals. The method involves taking rats, putting them in small cages, dropping mice into the cage and noting how long it takes the rats to kill the mice. If one rat kills a mouse in two minutes as opposed to another which takes four minutes, then the first rat can be claimed to be twice as aggressive as the second rat.

In order to complete the scientisation of what is going on, this 'mouse-killing' behaviour on the part of the rat is published in the scientific literature, and grants are given for a study of 'muricidal activity' (which makes it a great deal more scientific!). The way is then open to extrapolate this finding into the human situation, so that if a drug is found which affects the rate at which rats kill mice, this can be regarded as an 'anti-aggression drug'. (Drugs of this sort have now been clinically tried on 'schizophrenics' and what are described as 'violent hospitalised criminals'.) So the reductionist programme generates a range of technological outputs, techniques of knife, electrode and chemical for interfering with brains, and thus modifying behaviour.

But to return to the analysis of the flaws in reductionism. Arbitrary quantification is the ascription of numerical values to qualities which cannot in reality be adequately encapsulated in this sort of way. There follows what might be described as arbitrary algebraicisation; that is the belief that, if one can ascribe a number to some property, one must be able to partition out that property into a proportion given by nature and a proportion given by nurture, with a minor additive term for interaction. This nature-nurture dichotomy is another major example of the way in which reductionist thinking forces one into wholly spurious problems; the dialectic between interacting processes each of which modulates the other is turned by reductionism into an arbitrarily polarised contest between two static and separate, quantifiable forces.

Reductionism having emphasised that all human phenomena can be divided up into a category given by genes and a category given by environment, with very little scope for interaction between them, it is scarcely surprising that when the algebra is applied it turns out that everything, or a very large proportion of everything, is inherited, whether it is radicalism, intraversion, aggression, intelligence, twins' propensity to answer questionnaires consistently or inconsistently, or the ability to learn French at school. Such examples, drawn from heritability studies carried out over recent years, offer conclusions which have substantial social resonance and political significance, but are strictly empty of scientific content.

The final step in the sociobiological reductionist model is to argue that, if all the properties of societies are merely the properties of individual members of that society, and these properties are genetic and hence inherited, it should therefore be possible to find an adaptive evolutionary explanation for them. This gives rise to the whole pattern of adaptationist myths, of which sociobiology is the most important current example.

Such evolutionary myths for the origins of what are regarded as adaptive properties of individuals have become very popular. Just one example is the argument that the present division of labour between the sexes in contemporary Western society is given by an adaptation which derives from the human gatherer-hunter past, in which men went out and caught big animals whereas women stayed home and nurtured the children, and therefore men got genes for greater spatio-temporal ability whereas women got genes for linguistic ability. Hence, we are told, men are executives and women are secretaries.

The logical conclusion of this process is the Beatrix Potter syndrome, the attempt to project human qualities onto animals and then see the existence of such animal behaviours as reinforcing one's expectations of the 'naturalness' of the human condition. When one finds in the animal world exactly what one expects from the human world, one can then translate that observation back to the human world again so as to confirm it's 'naturalness' — children are naughty because Peter Rabbit behaves like a human child. This process is not new — it was described as long ago as the nineteenth-century by Engels in response to the social Darwinism of the period — but it is insidious, and its implications are spelt out more fully in the next chapter.

The processes outlined in this chapter are key steps in the methodology of reductionism in biology and show why it is both seductive and fallacious. Each of the steps in the argument has to be challenged, not merely because each generates dangerous technologies or poor ways of thinking, but because each contributes to preventing an understanding of the complex reality of the biological world.

Biological reductionism has developed, historically, until today it has become the major mode of explanation in biology. Reductionist explanations do more than merely sustain the ideology of the dominant class, race and gender, and generate obnoxious technologies; but are also fundamentally flawed as ways of describing, interpreting and predicting the world. That is, reductionism is at best a partial, at worst a misleading and fallacious, way of viewing the world. Reductionism, which began as liberatory, has like capitalism itself become oppressive. Until today it limits our understanding of the universe.

Notes and References

1. There are troubles with the term holistic. It has acquired numerous meanings during its history, rather as has the term 'alternative' (as in alternative medicine, alternative technology, etc.). But I don't want to get caught up in discussion of the differing nuances of holism here. I am *not* using the word as a sort of universal praise term, the good fairy antithesis of evil reductionism, nor does it have any mystical connotation here. On the contrary, I argue in this chapter and throughout the book that holism and reductionism are mirror-image types of explanation, and that what is ultimately needed is to transcend both, in a mode of description which may perhaps best be described as dialectical, or, if this word is considered to carry too much historical baggage, then perhaps 'integrationist'.
2. R. Dawkins, *The Blind Watchmaker*, Longmans, 1986, p. 111.

3 Less than Human Nature: Biology and the New Right *(with Hilary Rose)*

Politics and human nature theories

Embedded within any political programme and the theory which sustains it lies, implicitly or explicitly, a theory of human nature. Humans may be seen as innately competitive and xenophobic; innately co-operative; isolated monads acting in their own selfish interests; as playgrounds of good and evil; as fallen angels or risen apes; as infinitely malleable and shaped by the economic and social environment; or as cogs in the machinery of the iron laws of history. Whatever form it takes, it is the concept of human nature which provides the inner core of consistency to political theory.

Pragmatic politicians may disguise or fudge their theories, but political movements of any ideological significance need to express them clearly. For such movements, their concept of human nature cannot be *derived* from political theory but must appear as metapolitical and prior to the theory; it needs, therefore, an external source of legitimation. The source of the human nature concept may be god-ordained, as generally in Europe before the nineteenth-century, or for fundamentalist Christians or Islam today, or it might be a version of Enlightenment rationality. But increasingly from the seventeenth-century on and with particular force from the mid-nineteenth-century, a most important source of legitimation for the political theories of the industrialised world has been science. Science claims to provide objective knowledge of the organisation of the material world and hence, although it itself is regarded as value-free, its laws are taken as providing guidance as to how we may best live in accord with that material reality. Scientists and philosophers may claim that one cannot derive an ought from an is, but the history of the last three centuries is enough to reveal how often that precept, even were it to be accepted, has been broken. Initially following Newtonian physics, the rules of the cosmological and inanimate world were established;

increasingly following Darwin and the growth of evolutionary theory, the rules of the biological world too came to be asserted.

The consequences of this transition between religion and science as the ideological legitimator of the nineteenth-century are of course very familiar. If one accepts the conventional, but increasingly challenged account of the autonomy of scientific theory, then it is commonplace to recount how the metaphors of evolution, of the 'survival of the fittest' and the 'struggle for existence' made their way into Spencerian sociology and Social Darwinism, which found biology to be in support of *laissez faire* capitalism and class inequality at home and imperialism abroad. Marx and Engels were quick to point out of course that the reverse was also true; that *laissez faire* capitalism was on the side of Darwinism; that Darwin had looked at the non-human biological world through the distorting glasses of a typical Victorian gentleman. Nonetheless, they too found in Darwinism the legitimation both for a science of history and a progressivist vision of the evolution of human society through class struggle. This ambivalence of Marxism to science, at once praising and claiming to emulate its objectivity and finding ideological values embedded within it, has characterised much of the left's attitude to science in the subsequent century; typically the right has been less troubled.

Theorists of racial and sexual inequality too quickly found sustenance from science. Nancy Stepan[1] has shown how racism — that 'scavenger ideology' — seized on every version of nineteenth-century anthropology and theory of human origins, polygenist or monogenist, to demonstrate the 'truth' of white superiority. Elizabeth Fee and others[2] have shown a similar pattern of argument both shaping the research of the physiologists and the pronouncements of the patriarchs with respect to anything from the size of the female brain to the insatiable demands of her uterus. The Whig view of scientific and political progress tends to assume that we are now far removed from that vulgar past; whatever the failings of science then, today it is freer of values, and today's political theories more sophisticated. As we will see, however, this is far from the case. Of course we should not expect to find a precise correspondence between a person's class, gender or race and their scientific, ideological and political beliefs. Real life is a good deal messier than that. This is partly because ideas cannot simply be read off from some version of the economic 'base'; ideology takes on a certain life of its own, touching untidily the social interests it serves. Nor are individual actors at the mercy only of a single determination. It is this paradox which has led to such contortions by that school of historiography and sociology of science which has tried to read all scientific theories and experiments as 'merely' the products of the concatenation of 'interests' which any actor represents. At the same time, scientific theories are constrained by the need to bear some relationship to the material world they are attempting to explain. In the case of the racist anthropology of the nineteenth-century, both monogenist and polygenist theories represented scientific advances over the doctrines of special creation which preceded them; scientifically, it was a step forward to relate, or try to relate, differences in temperament or 'intellect' in humans to physiology rather than attribute them to God having created Eve from Adam's rib.

In this chapter, we want to show how the 'new synthesis' of sociobiology which emerged in the mid-1970s as the focal point for debates about the biology of human nature, has become central to the New Right's vision of human nature, aided and abetted by the pronouncements of the sociobiologists themselves. The political scene we analyse is that of Britain in the 1980s, but comparable accounts could be given of the New Right elsewhere in Europe — especially France, and in the USA.[3] While sociobiology makes a number of very broad claims about the human condition, here we will concentrate specifically on its emphases on the origins of race and gender inequalities and their corresponding political expression in racism and sexism.

Sociobiology's origins are simultaneously internal — arising from developments in ethology and evolutionary biology which set its agenda — and external. Assumptions about the appropriate way in which to define, reify and analyse the behaviour of human and non-human animals, which aspects of such behaviour are appropriate to observe, what types of mathematical treatment to apply to them, are clearly not neutral or value-free. These issues were apparent in the writings of those evangelists for sociobiology whose books came to dominate and define the field — E.O. Wilson, Robert Trivers and Richard Dawkins and the popularisers who soon followed them. After a decade of criticism from many quarters, radical and non-radical, from inside and outside the biological sciences, some of those working in the field which sociobiology claims as its own have endeavoured to distance themselves from the evangelists, and to rescue what they regard as a 'rational kernel' of sociobiological theory from its more transparently ideological accretions. This rational kernel will be discussed below. 'Sociobiologists' do not constitute a monolith, and we are not arguing that questions of 'why we do what we do' (to quote one of sociobiology's catch phrases) are illegitimate or can be divorced from questions of biology even if they are frequently wrongly posed — or even pure ideology masquerading as science.

It is this complexity which makes discussions of 'interests' in relation to scientific theory-making so fruitless. Sociobiologists are not necessarily racist, fascist or paid-up members of the New Right simply because their theories provide sustenance for inegalitarian political philosophies. The discussion is about the social functions of human nature theories and the social context in which they emerge, not about individual motives or even individual interests.

The left and human nature

On the traditional (and predominantly male) left there has been a century of debate over human nature. Maxism itself has been ambiguous. One strand has denied any essential human nature at all; it sees humans as entirely the products of social and economic circumstances, determined by a materialism which seems to exclude biology from consideration. Such a reading of Marx was given credence both by Stalin's version of dialectical materialism and by western European Trotskyism. By contrast the anarchist reading, following Kropotkin, has claimed an innate co-operativity among humans, a co-operativity derived from our biological natures and hence itself the product of Darwinian evolutionary mechanisms.[4] And a radical reading of Marx, typified by the writings of Lukacs, and more recently Heller,[5] has re-established the centrality of human needs as historically constructed, but nonetheless the result of a continuous dialectic between biology and society, at the heart of the Marxist programme.

In Britain, and within the Labour movement, probably up to and including the 1950s, political thinking was shaped by human nature theories derived from an uneasy amalgam of Marxism, Christian socialism and anarchism. Present political positions and programmes could be matched against a vision of socialism expressed by William Blake's New Jerusalem, or the xenophilic import of Stalin's New Communist Man [sic]. But through the 1960s and 1970s there were rapid changes. On the one hand a succession of Labour administrations, committed first to fueling consumerism and later to pragmatic monetarism, fudged and mudged away at the possibility of any socialist vision at all, so that by the mid-1970s it seemed doubtful if any Labour politician would recognise the new Jerusalem even if you pressed their noses to a plate-glass window looking out over vistas of England's green and pleasant lands. On the other hand, the New Communist Man, it was plain to all, turned out to be the dismal product of Gulag Archipelago, vodka and Russian Orthodox mysticism.

Into this economically still prosperous but politically moribund scene burst the student movement with its passionate utopianism, and then the new feminism. Hard on their heels came the sociologists whose theorising moved the new social movements away from the traditional *historical* relativism of the left in favour of a view of the world as infinitely plastic and socially constructed. Social constructionism dissolved away all the old-seeming certainties of human existence — from the dependence of infancy through to the frailty of old age and the existential despair of madness — into nothing but matters of social labelling, by those with power over those who had not. This was the age of Laing, of Foucault, of hyper-reflexivity.[6]

The result was a vacuum; there ceased to be criteria against which any one way of living, form of social organisation or piece of scientific data could be judged as better, more in accord with the material world, or more appropriate, than another. Male radicals, just as the male labour movement, were left without a vision of the new society or new possible human relations, without even a sketchmap of how to move beyond the present, other than the merest voluntarism, or withdrawal into isolated utopian experimentation in personal life-styles, increasingly detached from the lived experience of the majority of the population. (This is not the place to discuss the even more profound debates within feminism.)

Conservatism and human nature

The general vacuum left by such social constructionist theories made space for the rapid advance of human nature theories of the New Right in the later 1970s and in the 1980s. The post-1945 Conservative party had suffered a similar transmutation towards pragmatic management of a mixed economy and welfare state as had the Labour party, its residual rhetoric deriving from that patrician, feudal, one-nation strand of Toryism running from Disraeli to Macmillan. For some time the voices reasserting the old Adam, humanity's original Social-Darwinist nature, were to be found mainly outside the party, among the neo-Fascist and racist groupuscules, the National Front and its multitude of sectarian relatives. Their language and tradition is that of Social Darwinism transmuted by fascism and Nazism into an almost mystic but biologically derived concept of the race, its need for biological purity and protection from the dangers of race-mixing and, of course, its manifest destiny and mysterious shared values of blood and gene with white 'races' abroad.[7]

The 1985 policy statement on 'racial nationalism' by the National Front makes the point succinctly when it concludes:

> It has been proved conclusively that the average intellectual level of the white race is superior to that of most other peoples ... racialism, far from being an anti-social aberration is an instinct inherent in all peoples and all species.[8]

For many years almost the only voice of human nature rhetoric within Parliament was that of Enoch Powell. But as the 1970s wore on, the rhetoric was to be found more and more on the lips of the leading politicians of the Conservative Party. In the run-up to the 1979 election Margaret Thatcher could be heard proclaiming, as if in explicit counter-emphasis to the left's social constructionism, the 'naturalness' of competitiveness, of the desire to pass on inherited wealth to one's offspring and of racial loyalties against the fear of being 'swamped' by aliens. By the time of the first Thatcher government the concept of human nature in its most inevitabilist and conservative form was entrenched in the language of the party, 'wets' and 'dries' alike. It emphasised the unchangeability of human nature, the natural categories of the family, the sexual division of labour, the capitalist form of economic organisation and the racial integrity of the British people as apparently preordained by biology.

We will consider shortly the meaning of these uses of the concept of human nature and their biological grounding more systematically, but is worth starting by considering some representative voices. Here, for example, is Enoch Powell in the debate on the 1981 Nationality Bill, arguing that British citizenship should be passed on only through the father on the grounds that plans to let a child claim nationality though its mother were:

> a concession to a temporary fashion based on a shallow analysis of human nature ... Men and women have distinct social functions, with men as fighters and women responsible for creating and preserving life; societies can be destroyed by teaching themselves myths which are inconsistent with the nature of man [sic].[9]

And a now-familiar quotation from Patrick Jenkin, then Social Services Minister, on the biological inevitability of the division of labour:

> Quite frankly I don't think mothers have the same right to work as fathers. If the Lord had intended us to have equal rights to go to work, he wouldn't have created men and women. These are biological facts, young children do depend on their mothers.[10]

Each roots policy in a concept of human nature, and human nature in biology (though Jenkin effortlessly achieved the almost impossible task of invoking a *double* legitimation, of biology *and* God, for his assignation of gender roles).

As for competitiveness, property and ownership themselves, when Thatcher claims that:[11]

> We are all unequal. We believe that everyone has the right to be unequal

or that

> The charm of Britain has always been the ease with which one can move into the middle class. It has never been simply a matter of income, but of a whole attitude of life, a will to take responsibility for oneself

she is expressing a doctrine of the essential justice of inegalitarianism which derives from an explicit Social Darwinism and which underlies, as we shall see, both the libertarian and authoritarian forms of New Right philosophy.

It is perhaps not surprising that in such examples the concepts of *nature* and *naturalness* are used in a variety of different ways. The first of these might be called statistical, that is what most people actually do or are believed to do (want a family and property, resent immigrants, and so forth). The second sense is more prescriptive, it defines how things *ought* to be done. Discussions of the centrality of the bourgeois nuclear family are of this type; it is seen as a *natural* social unit because that is how the commentator thinks that society ought to be structured. If it is natural, that is, wholesome or 'good for you', then to establish or seek to defend social institutions which protect it is merely to act in accord with nature. And of course other social forms of the family or other collective units then become doubly *unnatural*; unnatural first in the sense of the *a priori* definition and then in relation to the social institutions of the majority.

The third sense of *natural* is to imply that something cannot be substantially changed because it is underwritten by human biology, fixed in our genes as a result of human evolution, and often something that we are seen as sharing with at least some other species. This is the sense in which we are told that 'you can't change human nature', and that it is dangerous to try. Yet these concepts of nature, and above all human nature, as unchangeable, merely requiring to be read off from some immutable unchanging biological reality, are themselves constantly subject to the relentless modernising drive of science and technology within a capitalist society. For instance, over the last few years, the issues of *in vitro* fertilisation and surrogate motherhood have laid dramatic siege to our sense of what is 'natural'.

The New Right

The point is that human nature is firmly back on the agenda as an explanation of our social ills and as a justification for social formations. And in part because the left has abandoned the concept of human nature, the vision we are offered is that of the right. It is not surprising that human nature is a term bandied about by politicians of the right from fascists to Tory wets, and that they use the term in a variety of different ways. But it is with the specific theoretical positions of the New Right that we are most concerned here, for these may be taken as the coherent cutting edge of an ideology which mainstream politicians wield more pragmatically.

For our purposes here we want to define two contrasting tendencies in New Right[12] thought. On the one hand, there are the neo-Liberals, committed to the centrality of the market mechanism, to deregulation, privatisation and the reduction of the power of the state. Their philosophical guru is the Austrian Friedrich von Hayek and in Britain they are grouped around the Adam Smith Institute.[13] The alternative strand, far from liberalising and deregulating, preaches the values of authoritarianism and the strong state. It decries liberals almost as vehemently as it does Marxists, and sees the function of government as to control humanity's baser instincts in favour of a higher morality. In Britain, this tendency is typified by aesthetics professor Roger Scruton, editor of the *Salisbury Review*, and author of a stream of books and essays on themes ranging from architecture to the nature of sexuality.[14]

Despite the apparent contrast between these two ideological positions, it is our argument here, (and this is a point which has been made by many others), is that they are but two faces of the same coin, minted in the long tradition of biologically determinist views about human nature. This is easy to see in the case of the libertarians, whose critique of the growth of state power rests upon the Hayekian tradition of methodological individualism, with its emphasis on the priority of the individual over the collective. That priority is seen as having both a moral aspect, in which the rights of the individual have absolute priority over the rights of the collectivity, and an ontological aspect, so that the collectivity is nothing more than the sum of the individuals which make it up.

Philosophically this view of human nature goes back to the emergence of bourgeois society in the seventeenth-century and to Hobbes's view of human existence as a war of all against all, leading to a state of human relations manifesting competitiveness, mutual fear and the desire for glory. For Hobbes, as for Hayek today, it followed that the purpose of social organisation was merely to regulate these inevitable features of the human condition. And his view of the human condition — even in those dim and far-off pre-sociobiological days — derived from his understanding of human biology. It was biological inevitability which made humans what they were; modern sociobiology improves on Hobbes by even deriving co-operation and altruism from innate competitive mechanisms.

The alternative, Scrutonian position, with its authoritarian emphases, seems at first sight removed from the Hobbesian tradition, and even Platonic in its view of humans as possessed of fixedly unequal qualities and the need for the masses to be regulated to the finest degree by an authoritarian state. It is, however, but the other side of the same biological coin; for if human existence is in the Hobbesian sense nasty, brutish and short when unregulated, then the differences between Scruton and Hayek become those of the degree of regulation necessary to control the brutishness. Paradoxically for Scruton it becomes the task of the state to uphold inequality and assert individualism. The frequent emphasis given to the concept of nature within Scruton's writing has a double meaning; it claims simultaneously to speak with the authority of science, and a privileged access into an almost mystical vision of human essentialism. This combination too, has a long lineage in right and extreme right thought.

The scientific elements within such writing are concerned with the apparently fixed nature of human needs, located in physiology and the irreducible nature of typological and individual human differences. The submerged themes are those of the origins of these differences in *genetic* differences, just as supposed human universals arise out of alleged genetic *similarities*, and the alleged evolutionary reasons for the fixation both of the universals and of the differences. This reductionist and biologically determinist combination characterises the New Right, as it did its ideological predecessors. It offers to explain social phenomena in terms of individual behaviour (for instance, societies are aggressive and wars occur because the individuals which compose them are aggressive). Both Hayek and Scruton employ the elements of the theory with a rather broad brush; while Hayek is concerned to offer an account of human evolution to which we will refer in more detail below, Scruton deals in *ex cathedra* assertion. It has been left to their popularisers to make the more specific links with sociobiological theory and hence to claim that the new science is in accord with social theory. A few examples may suffice:

> Science now seems to have caught up with Adam Smith. To support an economic lame-duck is not merely bad economics, but apparently against our deep-seated nature. It will be a chilling thought to many, but it does begin to look as though we have a little bit of Maggie Thatcher in us all.[15]

> Bioeconomics says that government programs that force individuals to be less competitive and selfish than they are genetically programmed to be are preordained to fail[16]

> If Professor Eysenk has been right all along, the implications for our educational, social and penal policies is enormous. For the hard fact is that if he's hit on the truth, then much of our society is based on a lie ... He says that in key areas of pay policy, education and dealing with crime, attempts to impose so-called 'equality' have been disastrous and that we're suffering from the inevitable backlash.[17]

The rise of biological determinism

As argued in earlier chapters, within biology itself there has been a long-lasting tradition of reductionism, the origins of which are intimately connected with the history of the birth of modern science in general and biology in particular in the context of the emergent bourgeois order of seventeenth-century Europe. The implications of this strand within the development of evolutionary theory, and later in statistics, eugenics and psychometry, has often been discussed, and may be summarised simply by saying that throughout the late nineteenth and early twentieth-centuries genetic and psychological theory-making, experiment and observation were bound up closely with proposals for public policy and with ideology, from the arguments that poverty ran in the genes to those calling for eugenic sterilisation of moral degenerates and for restriction of immigration of those of inferior intellect into the USA.[18] The determinist arguments and spurious data on which such theories and proposals were based were refuted in debate by geneticists, anthropologists and social scientists through the 1930s and swept away politically in the aftermath of the Nazi holocaust, leaving a consensus among the majority of natural and social scientists which lasted until the mid-1960s and which emphasised the importance of the environment — of nurture over nature — in the formation of human traits.

The re-emergence of biological determinism as a significant force both in scientific and ideological discourse can be conveniently dated by the appearance of Arthur Jensen's article in the *Harvard Educational Review* in 1969. The article, as with so many subsequent determinist writings, did two things simultaneously; it commented on an assumed social and political problem, and claimed scientific validity for a hypothesis to explain the origins of the problem. Jensen, an educational psychologist, was concerned about the relatively poorer performance of blacks than whites in schools in the USA. In the aftermath of Project Headstart, a compensatory education programme which was part of President Johnson's 'Great Society' initiative, Jensen claimed that the policy had failed, that social factors were not enough to account for observed black/white differences in IQ and that therefore there must be a genetic explanation. Some eighty per cent of the differences in intelligence between individuals was inherited and:

> there are intelligence genes, which are found in populations in different proportions, somewhat like the distribution of blood types. The number of intelligence genes seems lower, overall, in the black population than the white.[19]

Hence no programme of social action could equalise the social status of blacks and whites, and blacks ought better to be educated for the more mechanical tasks for which their genes predisposed them. The claim of genetic inferiority of blacks was rapidly extended to the working class in general by Harvard psychology professor Richard Herrnstein and incorporated into public policy discussions by President Nixon's advisers.

Jensen's claims crossed the Atlantic, with Hans Eysenck his former teacher (and Cyril Burt's former student). Eysenck has a long history of pronouncements and publications on the inheritance of almost anything from intraversion/extraversion to consistency in answering questionnaires, political attitudes and wealth. He concluded that:

> It seems certain that whenever blacks and whites are compared with respect to IQ, obvious differences in socio-economic status, education and similar factors do not affect the observed inferiority of the blacks very much ... this inferiority ... cannot be argued away as being due to lack of motivation.[20]

For Eysenck what was true for blacks was true also for the Irish and the working class in general. Eysenck has remained a tireless propagator of these hereditarian views, and is one of the few scientific figures referred to here who has apparently been without qualms about being quoted by journals of the far right, having given interviews to neo-fascist magazines and lending his name to the editorial board of the journal *Mankind Quarterly*. With or without the intent of its principal advocates, the claim of the genetic inferiority of the blacks or non-Europeans in general rapidly found its way into National Front leaflets and propaganda in Britain and abroad; the pronouncement from the National Front quoted above is characteristic in the claims it makes for 'scientific proof' of white superiority.

The claims of the inheritance of IQ and the origins of race differences were of course vigorously disputed. Much of the central evidence was based on the data of Cyril Burt, exposed as fraudulent by Leo Kamin[21] and by Oliver Gillie.[22] Kamin's systematic reassessment of the original data, and the attack on the theoretical basis for estimates of heritability and the significance of IQ measures, by Richard Lewontin in the USA and by others in Britain[23] went a long way towards defusing the issue. By the 1980s, court judgements in California had ruled against the use of IQ tests as being discriminatory; in Britain the massive Swann Report on the performance of ethnic minority children in schools, published in 1985, firmly ruled out any possibility of an explanation of 'under-achievement' based on genetic differences between races.[24] The issue runs as an open claim only in the writings of overtly fascist and racist groups; within the broader New Right, the language of inferiority has been replaced by coded references to 'difference' and to 'alien cultures', as when the leader of France's Front National, Le Pen, refers to the existence of:

> different races, ethnic and cultural groups in the world which ensures a distinction between people and nations ... I would not say that Bantus had the same ethnological aptitudes as Californians because this is simply contrary to reality ... it is evident that hierarchies, preferences and affinities exist.[25]

By this time the specific issue of genetic differences in intelligence had been subsumed within the much broader ambit of the claims by sociobiology that racism and nationalism were but natural extensions of tribalism, itself the product of a process of 'kin selection' which lies at the core of sociobiological theory (as in the National Front statement quoted above). Irrespective of whether other ethnic groups were superior, inferior or just different, we were genetically programmed to wish to avoid them. The most recent publication from Scruton's *Salisbury Review* group, makes exactly the same point. People seek out those for whom they have 'natural' affinities and similarities, in culture and ethnicity, and this human tendency is to be endorsed even while overt racism is to be disapproved of.[26]

Meanwhile the eugenist arguments of the 1920s and 1930s (dressed in only mildly refurbished clothes), that located social dissidence from crime to inner city riots in the biological realm, were all to be renewed (see Chapter 4). For the spectrum of politicians of the right, the emphasis is thus on faulty or maladapted genotypes in those weaker citizens unable to adapt to the challenges of an industrial society, genes which do not allow them to 'get on their bikes' in the search for jobs or to break out of the cycle of deprivation. The ideological point of such explanations is not the precise biological mechanism involved, but that *any* explanation which locates the cause of social distress in the biology of the individual serves as a victim-blaming mechanism essential to the methodological individualism of New Right philosophy. Such weaker genotypes must, if we are to preserve the Darwinian edge that social evolution demands, be allowed, with regret, to go to the wall. That is the law not of the jungle but of biological theory making, and it is offered not in a vindictive or punitive tone, nor even in the voice of the prophet denouncing original sin, but in the weary but civilised tones of scientific law-enforcers (who don't make the law, merely interpret it).

As the 1970s wore on, the ideological defence of the status quo, challenged by the rising claims of blacks and the social disintegration of the inner cities, found an even more fundamental challenge being mounted by the new feminism. The biological naturalness of the sex/gender divisions in society were a conspicuous feature of post-Darwinian biological research and the ideological construction of women in the nineteenth-century. Biologically based arguments insisted that women were less capable than men of study or intellectual pursuit as a consequence of deficiencies in brain structure, the possession of a uterus or of the irresistible pressure of hormones. Women's entry into higher education, suffrage and most areas of 'men's work' have been opposed on the grounds that they were constitutionally less capable than men for these roles. Such arguments, refined by the modern languages of neurobiology, physiology, endocrinology, psychology and genetics have been extensively refurbished in the last decade in clear response to feminism. While arguments for the 'inevitability of the patriarchy' have come in a number of guises,[27] here we concentrate on the sociobiological claims as to the naturalness and origins of the sex/gender divisions within society, and the uses made of them by New Right theorists.

The task of sociobiology has been to show that gender differences are *adaptive*, that is, they contribute to the continued evolutionary success of both sexes, and they emerge because of the different reproductive roles of the two sexes. Because for men the business of fathering takes but a few moments — what sociobiologists refer to as 'male investment' in reproduction — whereas for women mothering is a long drawn-out process, men have an interest in roving and spreading their seed about, while women have an interest in ensuring the sexual constancy of their partners. Especially because of the long gestational and dependent period of the child, a division of labour emerges in which, as E.O. Wilson puts it:

> in hunter-gatherer societies, men hunt and women stay at home. This strong bias persists in most agricultural and industrial societies and on that ground alone appears to have a genetic origin. My own guess is that the genetic bias is intense enough to cause substantial division of labour even in the most free and egalitarian societies.[28]

while for fellow sociobiologist David Barash:

> Waging war is (and probably was) almost an exclusively male pursuit in all human societies. This is not to say that men are therefore smarter than women but simply that their greater size, strength and aggressiveness — all perhaps the evolutionary result of engaging in warfare — have given the average man a competitive edge over the average woman.[29]

The sex-gender division of labour is thus the consequence of the adaptiveness of the biological division of labour, and genes fixed in human evolutionary past now ensure that men are executives while women are secretaries.

The first point to note about this type of analysis is that, as we should not perhaps be surprised to find, the majority of the claimed 'biological facts' are based on evidence which is faulty, of uncertain provenance or significance, or are simply assertions based upon speculation.[30] Such 'just-so' fairy tales of human evolutionary past, beloved of sociobiology — and indeed of Hayek — are just that. The recent painstaking reanalysis of the anthropological and human prehistoric record by a new generation of feminist scholars makes this clear.[31] But from our present point of view, contesting the data and theories offered by sociobiology is not the main point; rather we want to emphasise the ideologically organising role played by their interpretations of the data.

The quotations from Barash and Wilson are exemplary in this regard. When the first volume of E.O. Wilson's trilogy, called *Sociobiology*, first appeared in 1975 with the full weight of Harvard University Press's publicity machine behind it, Wilson was careful to distinguish his approach from the work of ethologists like Ardrey, Morris and Lorenz who had preceded him into the business of telling us 'why we do what we do' (to quote a film made by another leading sociobiologist, Trivers). Theirs, Wilson explained were works of advocacy. His was a work of scholarship. (In similar vein, when sociobiologist Richard Dawkins was informed that his book *The Selfish Gene* was being used as the basis for racist propaganda, and asked to dissociate himself from the New National Front, which had quoted him, he judged it sufficient to respond that his work was above 'the ephemeral level of human politics'.[32] Yet it is clear from the extracts from Barash and Wilson, and from others that follow below, that this is far from being the case. Characteristically, such sociobiologists adopt a double persona; while claiming the objectivity of the scientist whose motivations and theories are neutral and above suspicion, they also cloak social and political nostra in flights of ideological fantasy which are certainly not readily distinguishable from what the observer might *experience* as advocacy. Their views are made all the more persuasive precisely because they are offered as science and therefore carry the cachet of 'objectivity'.

However, whatever the intent of the theorists, by the time claims for the biological bases of patriarchy have reached popular and political presentation they lose any sense of being provisional scientific observations. They are frozen into inevitability; the biological construction of gender differences becomes mere scientifically proven common sense. One newspaper report, for example, quoted the 'considerable displeasure' expressed by the women's movement over research which suggested that 'sexism ... is a natural, inevitable and permanent feature of personality' and which challenges

> the view that sexual stereotypes are an invention of a male-dominated society and indicate instead that they arise, in part, from a biological programme, the legacy of millions of years of evolution which starts to unfold at birth.[33]

The argument of this article was that feminist claims that gender-specific behaviour is environmentally conditioned are refuted by 'scientific evidence' which is said to prove that gender differences in behaviour are biologically determined. A similar argument was made by Eysenck's colleague Glenn Wilson, who specialises in writing about what he calls 'the sociobiology of sex differences', when he claimed that the hegemony of social constructionist theories has led to a feminist 'witch hunt' which disallows alternatives. He goes on to argue that there is already greater equality between the genders than feminists admit, because 'men and women are equal to the extent to which they are predisposed by their biological natures to behave in particular ways'. Finally, like others who disavow advocacy, he raises the question of the social policy implications of his determinism, concluding that 'there is a limit to the extent to which the feminist movement can override the natural inclinations of men and women by persuasion and political power'.[34]

A characteristic trick of many authors in this area is that of seeing themselves as Galileos, intent on bringing scientific truth to light in the teeth of powerful opposition. They portray the feminist, anti-racist and socialist critique of a sexist, racist and hierarchical society as if these critiques were manifestations of the dominant ideology rather than the reverse. But far from being the heroic minority scientific opposition of a Galileo or a Bruno, the new biological determinists claim their support from today's church and inquisition.

The policy implications of genetic determinism in this context are also considered by E.O. Wilson in the second part of his trilogy, *On Human Nature*.[35] He outlines three types of policy which a society might adopt in response to the knowledge of biological constraints on sex/gender equality. First, it can condition its members so as to exaggerate gender differences; second, it can provide equal opportunities but do nothing else; and third, it could train its members to eliminate such differences. Wilson regards the first option as providing a society 'richer in spirit, more diversified and even more productive than a unisex society' which could safeguard human rights while channelling men and women into different occupations. The second option would merely ensure that biology will out; men would be 'likely to maintain disproportionate representation in political life, business and science.[36] The third option would 'certainly place some personal freedom in jeopardy'. The assumptions are clearly both that sex-gender differences are desirable *and* that existing society offers women personal freedom. In any event, whatever we might wish to do to change gender inequalities, we will be constrained by the dictates of biology.

Insofar as these views are in accord with 'common-sense' beliefs in the naturalness of existing gender categories, it is not easy to separate out the extent to which sociobiology has directly influenced the thinking of the New Right, or whether the coincidence in views with those of politicians and ideologists is merely a reflection of a wider *zeitgeist*, The *kinder-kuche-kirche* view of women's place appears strongly in fascist writing as it did in the 1930s; it appears in Powell and Scruton. National Front publications are, as we have seen, particularly concerned to quote from scientific authority:

> One has only to observe the degree to which male dominance and female passivity in sexual courtship obtains in the animal world, likewise qualities of male aggression and female domesticity, to understand their fundamental biological basis. Such observations quickly demonstrate that 'feminist' talk of sexual roles being conditioned by society itself is the most puerile Marxist rubbish. Sexual and other behaviour differs between men and women simply because of differences in male and female hormone secretions which are governed by the sex chromosomes of our genes [sic].[37]

Roger Scruton's diatribe against feminism is similarly based on biology and waxes lyrical about male sexuality which he sees as by nature highly directed and scarcely controllable. Men are:

> The victims of an impulse which ... is one of the most destructive of human urges, and the true cause of rape, obscenity and lust.[38]

For Scruton, the biological potential of reproduction is 'what it means to be a woman', who cannot be defined otherwise. Although in his book on sexual desire[39] he argues against the reduction of human desire to 'mere' animality, to be interpreted sociobiologically, he goes on to reconstruct virtually all the sociobiological assertions about the propriety of bourgeois sexual behaviour, rooting them in *ex cathedra* statements about what it means to be a human being. But where does Scruton derive his authority for such assertions about human nature? We are never told; either they spring fully formed from his head, or we must assume that his disclaimer of the 'science of sex' is more of form than substance.

As well as confirming women's domestic place in the sphere of reproduction rather than production, sociobiology also offers guidance as to the relations between the sexes. Reviewing the evidence, Wilson finds that men are 'naturally' polygamous, whereas women are monogamous; Dawkins goes on to speculate as to how genes 'for' male philandering might be advantageous to the male and therefore spread through a population. The argument is based on the assumption, referred to above, that because females have a greater reproductive 'investment' in their young than do males — sometimes argued on the relative size of egg and sperm cells, sometimes on mammalian viviparity — then it pays them to care for each individual offspring, whereas it aids the spread of the male's genes if he 'invests' less in each individual offspring he has sired but endeavours to sire as many as possible. Equally, as only a female can be sure that the young she bears have her genes, a male must be jealous of 'his' female's chastity, whereas it is a matter of relative indifference to the female that the male be constant except insofar as he can be persuaded or cajoled into participating in the rearing process.

The argument extends from philandering to forced sexual intercourse. A characteristic of sociobiological discourse discussed in Chapter 2 is the wholesale adoption of the language of human social relations and its projection onto the non-human animal world. Thus we find reference to *harems* in baboons, *prostitution* in humming-birds, and *gang rape* in mallard ducks and even dung beetles (to refer only to the terms descriptive of gender relations; there is also slave-making and propaganda among ants, politics in chimpanzees, and many others). No longer is it now nature red in tooth and claw, but nature as adjunct to *Playboy*, which has, needless to say, gleefully taken sociobiology to its bosom. It is interesting that *Animal Behaviour*, the premier scientific journal in the field, eventually bowed to pressure from women ethologists and banned the use of the word 'rape' to describe what some human observers interpret as forced sexual intercourse among non-human animals.

For sociobiology, such acts, whether or not designated as rape, are the consequence of the male goal of maximising his chances of reproducing by mating with as many females as possible. For sociobiologist David Barash, even if human rape is 'by no means so simple as' what goes on between ducks, it is still squarely in the biological camp:

> Perhaps human rapists, in their own criminally misguided way, are doing the best they can to maximise their fitness ... Another point: Whether they like to admit it or not, many human males are stimulated by the idea of rape. This does not make them rapists, but it does give them something in common with mallards.[40]

It is not clear how Barash knows what turns on men who don't like to admit they are stimulated by rape, and *a fortiori* how he knows what turns on mallard drakes, but such poetic licence is not an uncommon feature of sociobiologising of this sort. It is worth comparing Barash's account with Scruton's view that rape is the inevitable result of the sexual impulse of which men are the innocent victims. For Scruton the cause of rape is 'the lust that seeks ... to relieve itself upon her body', if that is, it is 'left to itself' which it will not be 'if a woman has anything to do with it'. Scruton appears not to understand not only that it is usually women who are raped but that they would undoubtedly prefer that male lust were left unrelieved in such situations. Again, it is hard to see Scruton's position as in any way different from that of the sociobiologists. What is clear is that at the core of Scruton's view of gender relations are the traditional concepts of the double standard and of the madonna/whore image of women; and they also form the masculinist hidden agenda of sociobiological fantasising about the nature not merely of human females but also the females of other species.

As should by now be clear, the claims of sociobiology go beyond their recruitment in the interests of defining the inevitability of the patriarchy, or defending the view that racialism and nationalism are encoded in our selfish genes. The aim is to encompass all human societies into an explanatory framework based on evolutionary biology. Nor are Wilson and Dawkins content merely to describe the world; as we have said, policy conclusions flow from their pens. True, Wilson is cautious:

> If the decision is taken to mold cultures to fit the requirements of the ecological steady state, some behaviors can be altered experimentally without emotional damage or loss in creativity. Others cannot ... A genetically accurate and completely fair code of ethics must also wait ... We do not know how many of the most valued qualities are linked genetically to the more obsolete, destructive ones. Co-operativeness towards groupmates might be coupled with aggressivity towards strangers, creativeness with a desire to own and dominate ... If the planned society ... were to deliberately steer its members past the stresses and conflicts that once gave the destructive phenotypes their Darwinian edge, the other phenotypes might dwindle with them. In this, the ultimate genetic sense, social control would rob man of his humanity.[41]

Dawkins too, finds it appropriate to offer warnings of evolutionary doom awaiting those who transgress Darwinian imperatives (oddly, perhaps, for a man above mere human politics). If the following has echoes which are unanticipated in the writings of a zoologist whose main prior research had been into cricket song, their implications are nonetheless familiar. In *The Selfish Gene* Dawkins criticises the 'unnatural' welfare state where:

> we [sic] have abolished the family as a unit of economic self-sufficiency and substituted the state. But the privilege of guaranteed support for children should not be abused ... Individual humans who have more children than they are capable of raising are probably too ignorant in most cases to be accused of conscious malevolent exploitation. Powerful institutions and leaders who deliberately encourage them to do so seem to me to be less free from suspicion.[42]

This is strong language for books which eschew advocacy and lay claims to scientific objectivity (strange language too for a Darwinist in Dawkins's case, given that much of the rest of his text is a hymn to the evolutionary merits of spreading one's genes as far and wide as possible).

The relationship between such biological claims and the philosophical stance of both wings of the New Right should by now begin to be becoming apparent. They combine a Hobbesian/Hayekian individualism with dark hints that some sort of a morally authoritarian state is required that can ensure genetic fairness, and can legislate to avoid us losing our Darwinian edge or even to avoid the undue reproduction of undesirable genotypes. What is more, it is a clear legitimation for the status quo, a panglossian paradigm marred only by the repeated hints that the decline in Social Darwinian, Victorian values may already have resulted in some loss of fitness.

But such arguments go further. If present social arrangements are the inevitable consequence of the human genotype, then nothing can really be changed. As Wilson puts it:

> the genetic bias is intense enough to cause a substantial division of labor even in the most free and most egalitarian of future societies ... Even with identical education and equal access to all professions, men are likely to play a disproportionate role in political life, business and science.[43]

Small wonder therefore that sociobiology was soon called into play to offer a 'genetic defense of the free market' and that economists were offering parallels between the mathematical formulations of the sociobiologists and monetarism.[44]

Sociobiologists themselves declare a desire to avoid these political inferences and simultaneously seem actively to court them. Thus David Barash cautions:

> Concern has been expressed that human sociobiology represents racism in disguise: This is simply not true. Sociobiology deals with human universals that are presumed to hold cross-culturally and therefore cross-racially as well. What better *antidote* for racism than such emphasis on the behavioural commonality of our single species.[45]

Unfortunately, however, it turns out that to sociobiologists such racism *may* be the product of one of these assumed universals. According to Wilson:

> Nationalism and racism are the culturally nurtured outgrowths of simple tribalism.[46]

It isn't then surprising that the National Front can conclude that there is a:

> basic instinct common to all species to identify only with one's like group; to in-breed and to shun outbreeding. In human society this instinct is *racial* ... The great question of our time seems to be whether European man, the pinnacle of evolution, will destroy — through the unnatural notions which are the modern product of his intellect — what his inherited instincts have striven through those eons of time to preserve.[47]

The same theme echoes, in somewhat more coded terms, through the pages of Scruton's *Salisbury Review*.[48] It will be clear that in such speculations, the method of sociobiology is, in one interesting sense, to turn Marx on his head; where Marx was to emphasise that the key to the past lies in the present, sociobiology's emphasis on evolution explains the present in terms of a hypothesised past in which genetic traits for particular behaviours were favoured and therefore became fixed in the population. Hence the enthusiasm for evolutionary just-so stories, an enthusiasm shared with Hayek, who is committed to a demonstration that capitalist social forms evolved from earlier and more primitive 'tribal socialism'.

There are two points to be made here. The first is methodological. Both Hayek and these sociobiologists are concerned to draw parallels between changing human societies ('social evolution') and the biological processes of evolution. The parallels they draw, however, involve fatal ambiguities of mechanism. On occasion, the evolution of human societies is supposed to occur by specific genetic mechanisms — for example the alleged fixation of sex-linked genes for spatio-temporal perception in the male in human hunter-gatherer past. On the other occasions, however, what is being invoked is *analogy*. In a less determinist past, Darwinists such as Julian Huxley argued that in humans social evolution had in large measure superseded biological evolution as mode of change in human populations. Dawkins offers instead a formal parallel procedure. Just as he believes that evolution in the biological sense is about the change of populations of isolated but self-replicating competitive units called *genes*, so, he argues, the components of human culture can be reduced to units called memes, which also self-replicate and compete for survival. A selection mechanism similar to that by which records arrive in and then fall out of the top ten determines the fate of such memes. Dawkins' model has found little favour with other sociobiologists but has crept into anthropological literature. By contrast, Wilson offers a heavily mathematicised proposal for the *co-evolution* of culture and biology, formalised in the least well received of his trilogy, *Genes, Mind and Culture*.[49]

When Hayek discusses human evolution,[50] he uses the term in a sense which may be read as Wilsonian or as analogic. Primitive societies were socialist, he argues, and today socialism represents an atavistic throwback to these early forms. The development of hunter-gatherer societies however, produced a decisive break, in which the (male, inevitably) heroes were those who first initiated barter and exchange. From this grew the patriarchal family and the evolution of the morality of the market place through a process analogous to natural selection. (Here Hayek performs, for precisely contrary ends, an act of myth-making comparable to that of Engels on the origin of the family, and Sohn-Rethel in his derivation of abstract thought from the processes of exchange.)

But this Hayekian myth-making exemplifies the second point to be made here concerning the uses of evolutionary metaphors, and that is that in all probability his just-so story simply isn't true. The man-the-hunter version of the motor of human evolution has been sharply questioned by modern anthropologists, especially those working in an explicitly feminist framework.[51] They have made the point that gathering rather than hunting was likely to be the main source of human food requirements; insofar as this was a predominantly female activity, an alternative myth is equally possible in which females selected males who were not dominant and competitive but co-operative and nurturant, and therefore more likely to participate in child-rearing.

We are not arguing, of course, that Hayek or Scruton or other New Right ideologues were driven to their conclusions about the ordering of human society by reading sociobiology. While they may have read Wilson, and National Front theoreticians clearly have, Hayek's writings predate modern sociobiology in many respects and Scruton's views derive from a philosophical rather than an explicitly scientific tradition. The point is different, and perhaps more interesting. It is that sociobiology and the New Right share a common methodology, a reductionism which is specifically Hobbesian. Thus, sociobiology seeks the causes of social phenomena as located in the properties of the individuals who comprise society, and the properties of those individuals as located in genetic imperatives. Wilson's programme as laid down in *Sociobiology* is indeed reminiscent of Hobbes's *Leviathan*; to explain and prescribe for the entire human condition beginning with a few basic principles. Sociobiology's Hobbesianism arrives via Darwinism, and for both Hobbes and Darwin competition occurs not as a result of some fundamental property of organisms but as the inevitable consequence of the automatic self-reproduction of a machine-organism in a world of finite resource. It is this derivation which New Right philosophy shares with sociobiology.

The problems with the method of sociobiology lie precisely within this reductionism, even if we ignore the extent to which weighty conclusions and political nostra are mounted on the shakiest of evidence and often on no evidence at all. The methodological confluence is most strikingly apparent when we consider the theoretical constructs of sociobiology, its language system and its modes of mathematical analysis. Thus sociobiologists observing the behaviour of non-human species apply to it *game-theoretical* analysis; they regard animals as adopting *investment strategies* designed to *maximise rewards* based on *cost-benefit analysis* and *optimisation procedures* in relation to *time-budgets* for *optimal foraging strategies*. Whereas it might be argued (although we would not, Dawkins certainly *has* endeavoured to defend his language in this way[52]) that the use of terms like 'rape' or 'slave-making' to describe particular types of animal behaviour is 'merely' punning, the terms identified above are central to sociobiological theory-making, and the mathematical formulations adopted are formally parallel to those used in reductionist economic theories.

It is not surprising that monetarist economists find that sociobiology offers them biological sustenance for free-market theories, and that military strategists note that *arms-races* are to be found in the animal world too, because these phenomena were placed there by sociobiologists who derived their theories, not fallen from the sky, but from the very real human social world around them. The trick, of course, is to look at the biological world through lenses distorted by social expectations, and then to turn those back on the social world once more, to find that one's social theories are, unsurprisingly, sustained by the biology. In this type of sociobiological theorising and New Right philosophy we see two mutually sustaining sets of beliefs.

The reductionist methodology of sociobiology, like that of the New Right, decomposes social phenomena into the aggregates of the properties of the individuals which compose that society, using methods of reification. But it goes further by arguing that the properties of individuals are determined by their genes, and that genes 'act in their own interests', which are those of self-preservation and self-replication. This is why for Dawkins we become merely lumbering robots programmed by our genes, with our brains as 'on-board computers'.[53] What the vigour of this language is intended to convey is that in Dawkins' view Darwinian selection acts only on the genes, and that therefore behaviour, human or non-human, is merely the gene's way of making another gene, a copy of itself.

Dawkins and other sociobiologists believe this because, unlike other biologists, they are committed to the view that selection acts *only* at the level of the gene. A less reductive, and more appropriate way of viewing the processes of selection is to see them as operating at multiple levels, the levels of gene, genotype (the ensemble of genes each person possesses), phenotype (the organism generated by the interpenetration of genes and environment during development), and populations of organisms. However, once one is committed to the gene as the unit of selection, it follows that each gene is in some sense 'competing' with all other genes for replicative success, and the whole panoply of determinism follows. Sociobiologists never permit the existence either of multiple levels of analysis or complex modes of determination. They find it hard to see that even on their own model of competition between individual genes, it may well be in the most successful gene's interest (see how easy it is to use the language if one slips one's mind out of gear for a moment) to produce a phenotype which behaves co-operatively.[54]

The problem of co-operative behaviour among organisms has proved to be of the greatest complexity for sociobiology, for if each gene competes with all others, how can self-sacrificing or altruistic behaviour evolve? For sociobiology the answer lies in the observation that any individual shares genes in common with its brothers and sisters (50 per cent), and in varying degrees with cousins and other relatives. Hence a behaviour which might result in the destruction of the individual would be genetically advantageous, and spread by evolutionary processes, if it resulted in the saving of the lives, and hence the propagation of the genes, of close relatives. This is the phenomenon of 'kin selection' first proposed by J.B.S. Haldane and given mathematical precision by William Hamilton in 1964.[55] Kin selection instantly became the central mode of explaining aspects of social behaviour which had hitherto puzzled evolutionary biologists because they appeared to be to the individual's disadvantage albeit to others' advantage (like the alarm-call of a bird which spots a kestrel, thereby drawing attention and danger to itself but possibly protecting others). The alternative to kin selection, that the behaviour involved activity 'for the good of the species' (so-called group selection), was incompatible with Darwinian mechanisms which involved selection at the level of the gene.

At once it became clear that altruistic behaviour in humans — neighbours risking their lives to save a child from a burning house, the solidarity of striking miners or of a political movement — was nothing more than a perverse extension of the principle of kin selection. Of course, it could be argued that the neighbour or miner was calculating that self-sacrifice now would be rewarded by matching self-sacrifice later, in which case they would be demonstrating what sociobiologist Trivers called 'reciprocal altruism', which was merely enlightened self-interest. It also became clear why we had a genetic imperative to support our kith and kin, and reject foreigners, for they not merely have an alien culture, but alien genes too.

Thatcher, sociobiology would argue, was indeed obeying an evolutionary imperative when she defended 'our' kith and kin in the Falklands against the Argentinians. However, she failed to calculate the number of our kith and kin who could be allowed to be killed in this exercise before the enlightened self-interest of the genes said enough was enough, just as she forgot — a sad genetic slip — that the dead British soldiers shared genes in common with the Welsh-descended Patagonians too. It would be inappropriate to read too close a correspondence between the rise of kin selection theory within ethology and sociobiology and the replacement of collectivist with familial and essentially self-centred modes of social organisation at the centre of the political stage, both in Britain and the USA. But since these are the two countries in which sociobiological theorising has taken strongest scientific and ideological root, the parallel is bound to be fascinating to anyone concerned with the social processes of science.

There is one final element in sociobiological method which must be added to complete the picture. Darwinian evolutionary processes are concerned with the phenomenon which, following Darwin himself, is described as fitness. Evolutionary change occurs because like begets like (with minor variations), and all organisms produce more offspring than can survive to reproduce in their turn. This means that those offspring which do survive to breed are likely to be those which are better adapted, or more fit in some way to their environment. The offspring of these more fit individuals are likely also to carry the trait which is the 'fitter' one, and hence in due course it will spread through the population. (This is why a central paradox of Darwinism is that it is very good at explaining how organisms get better at doing their thing, but rather poor at explaining how new species that do rather different things can emerge, but that is another story, see Chapter 4.) The issue of present concern becomes that of defining which of those multiple features of an organism's phenotype are the adaptive ones on which evolutionary pressures can act. About this, as about human evolutionary origins, a host of just-so stories has grown up.

In this context, the tendency of sociobiology is to argue that whenever we see a common feature of the biological world around us, it must be adaptive; that is, it must be particularly well-fitted to serve a particular function and further, that there must be genes for it, enabling it to preserve its form or even improve it across generations. In this sense, sociobiology is the equivalent of functionalism in sociological theory, and once more there is a striking parallel between the biological and sociological modes of explanation.

It is this feature of sociobiology, the so-called adaptationist myth, which has laid it so open to the charge of panglossianism. If we live in an unequal society, then this feature of our social organisation must be adaptive and the product of natural selection. It is here that the biological *metaphor* for the evolution of human societies becomes transmuted into a biological *mechanism*. If a trait exists and is adaptive, then it must also be genetically determined; hence, there are genes 'for' all those features of our social organisation to which sociobiology claims privileged access, from xenophobia to philandering. It's all in the *gene*'s interests, even if we don't like the consequences. The trouble is that the grounds for arguing that a particular observed feature of behaviour, or indeed of physiology, is actually adaptive and therefore has been selected for are very obscure. A particular trait may be contingent, the result of historical accident and the 'tinkering' nature of evolution in which change in the organism is always a response to present and past environmental contingencies but has no way of plotting the future of the environment (if it had, the dinosaurs would never have gone extinct). Or it may be the inevitable consequence of structural constraints, or of some quite other feature which *has* been selected for.

The conclusions from this discussion of human nature theories, sociobiology and the New Right are to be drawn in several different frames of discourse. For biologists, the inadequacy of sociobiological theorising and its highly ideological commitments is a salutory lesson. It points to the failure of the reductionist programme within biology, and the necessity to develop modes of analysis of complex biological phenomena which do justice to their complexity, their rich interconnections and their multi-layered degrees of order. The phenomena we study do not occur in isolation in a frozen moment of time, and they cannot be understood divorced from the fact that they are simultaneously process and product, history and structure. Because we lack the conceptual tools to approach biology in this way, or at least the interesting biological problems, it is all the more important to beware of premature closure. The goal of Wilsonian sociobiology, rephrased in less triumphalist and imperialising language, is that of unifying the knowledge of the social and the biological — a goal which must be applauded. But until the modes of thinking that this synthesis requires are developed, we cannot advance towards it. They must be more holistic and less reductionist, more subjective and less objective, more dialectical and less static and mechanical.

For students of the sociology of science the 'case' of sociobiology is irresistible. It has extraordinary interconnections in methodology and conceptual structure with particular philosophical, sociological and economic traditions. Also fascinating is the explanation of the historical conjuncture which brought this way of thinking, dormant for many years, simultaneously into the centre of the stage in political philosophy, economics and biology. Endless Ph.D. theses are doubtless there to be written (if we survive that long) and those of us who have been and are protagonists in the debate are going to have to get used to being the objects of someone else's academic study.

For those concerned to counter the ideology of the New Right, exposing the roots of its human nature theories and cutting away the apparently solid biological grounding in which they are embedded is part of an important political struggle; not because if we show that the New Right is in error, it will *thereby* collapse, but because the ideological arena is one part, and an important part of, the terrain of struggle. But perhaps most importantly, for those of us whose overriding concern is the building of a new world in which humans can live harmoniously with their own natures, with other people and with the biological and natural world around us, we must draw conclusions from the collapse of both the traditional left visions of human nature and the social constructionism which briefly succeeded it. The feminist project, perhaps because it is harder for women to either deny or overstress their own biology, perhaps because of the nature of the reproductive labour which the social division of labour allocates to women, is likely to provide us with the starting point for such a new vision.

Notes and References

1. N. Stepan, *The Idea of Race in Science*, Macmillan, 1982.
2. E. Fee, 'Science and the woman problem: historical perspectives', in M.S. Teitelbaum (ed.), *Sex differences: social and biological perspectives*, Anchor-Doubleday, 1976, pp. 173–221. S.S. Mosdale, 'Science corrupted: Victorian biologists consider the woman question', *Journal of Historical Biology*, 11, 1987, pp. 1–55. J. Sayers, *Biological Politics: Feminist and Anti-Feminist Perspectives*, Tavistock, 1982.
3. European Parliament Commission on Racism and Fascism in Europe, 1986.
4. P.P.G. Bateson, 'Sociobiology and human politics', in S. Rose and L. Appignanesi (eds.), *Science and Beyond*, Blackwell, 1986, pp. 79–99.

5. A. Heller, *The Concept of Need in Marx*, Allison & Busby, 1977.
6. H. Rose, 'Hyper-reflexivity — A new danger for the counter-movements', in H. Nowotny & H. Rose (eds.), *Counter-movements in the Sciences*, Reidel, 1979, pp. 277–90.
7. See the regular analyses of fascist literature in the monthly anti-fascist magazine *Searchlight*.
8. Quoted in the *Guardian* 12 April 1985.
9. Quoted in the *Sun* 18 February 1981.
10. Mr Jenkin said this in a TV programme on working women. Quoted in H. Rose and S. Rose, 'Moving Right out of Welfare — and the Way Back', *Critical Social Policy*, 2 (1), 1982, pp. 7–18.
11. quoted *ibid.*
12. R. Levitas (ed.), *The Ideology of the The New Right*, Polity Press, 1986.
13. See for example their magnum opus *The Omega Report*, Adam Smith Institute, 1986.
14. In this context the most relevant is R. Scruton, *The Meaning of Conservatism*, Penguin, 1980.
15. *Daily Mail* 1978, quoted in M. Barker, *The New Racism*, Junction Books, 1981 p. 160.
16. *Business Week*, 10 April 1978, pp. 100, 104.
17. J. McLoughlin, 'Equality stops here, now', *Daily Mail*, 21 May 1977.
18. There are many histories of this period. See, e.g. R. Hofstadter, *Social Darwinism in American Thought*, Braziller, 1959; D.D. Pickens, *Eugenics and the Progressives*, Vanderbilt UP, 1968; K.M. Ludmerer, *Genetics and American Society*, Johns Hopkins Press, 1972; S. Chorover, *From Genesis to Genocide*, 1979; B. Evans and B. Waites, *IQ and Mental Testing*, Macmillan, 1981, D.J. Kevles, *In the Name of Eugenics*, Penguin, 1986; and N. Stepan, *op. cit.*
19. A.R. Jensen, 'How much can we boost IQ and scholastic achievement?', *Harvard Educational Review*, 39, 1969, 1–123.
20. H.J. Eysenck, *Race, Intelligence and Education*, Temple Smith, 1971.
21. L.J. Kamin, *The Science and Politics of IQ*, Penguin, 1976.
22. O. Gillie, *Sunday Times*, 24 October 1976.
23. L.J. Kamin, op. cit.; D.L. Eckberg, *Intelligence and Race*, Praeger, 1979; J.M. Blum, *Pseudoscience and Mental Ability*, Monthly Review Press, 1978; Science for the People Collective: *Biology as a Social Weapon*, Burgess, 1977.
24. *Education for All: The Report of the Committee of Inquiry into the Education of Children from Ethnic Minority Groups*, (Swann Report) London, HMSO, Cmnd 9453, 1985.
25. J.M. Le Pen, *Les Francois d'Abord*, 1984, pp. 167–8. See also P. Taguieff, 'Le Retournement du Gramscisme', in *Politique Aujourd'hui*, No. 1, July 1983, pp. 75–92.
26. F. Palmer (ed.), *Anti Racism: an assault on education and value*, Sherwood Press, 1986. This text uses the coded concept of 'a preference for one's own kind' in several of the essays, notably that by ex-Bradford headteacher Ray Honeyford. For Scruton the preference is for British 'high culture' which it is not merely natural but also right to prefer because it is the best by virtue of its past traditions and assimilation of Graeco-Roman cultural values. More down-to-earth is a sociobiological text well-regarded by *Salisbury Review* writer, palaeontologist Beverley Halstead. Entitled *The Sociobiology of Ethnocentrism* (eds V. Reynolds, VSE Falqer & I. Vine, Croom Helm 1987), it claims that there is evidence for 'a mild genetic tendency' to xenophobia!
27. S. Goldberg, *The Inevitability of Patriarchy*, Morrow, 1974.
28. E.O. Wilson, 'Human Decency is animal', *New York Times Magazine*, 12 October 1975, pp. 38–50.
29. D. Barash, *Sociobiology: the whisperings within*, Souvenir Press, 1981, p. 187.
30. See, for example, R. Hubbard, M.S. Henifin and B. Fried (eds.), *Women Look at Biology Looking at Women*, Schenkman, 1979; R. Hubbard and M. Lowe (eds.) *Genes and Gender II*, Gordian Press, 1979; H. Fairweather, 'Sex Differences in Cognition', *Cognition*, 4, 1976, pp. 31–280; K.F. Dyer, *Challenging the Men: Women in Sport*, University of Queensland Press, 1982.
31. S.B. Hrdy *The Woman that Never Evolved*, Harvard University Press, 1981; N.M. Tanner, *On becoming Human*, Cambridge University Press, 1981; L. Liebowitz, *Females, Males, Families: A Biosocial Approach*, Duxbury Press, 1978.
32. R. Dawkins, *Nature*, 289, 1981, p. 528.
33. J. Campbell, 'Happy to be prisoners of gender', *Evening Standard*, 7 December, 1983.
34. G. Wilson, 'The sociobiology of sex differences', *Bulletin of the British Psychological Society*, 32, 1979, pp. 350–3.
35. E.O. Wilson, *On Human Nature*, Harvard University Press, 1978.
36. Wilson, pp. 137–9.
37. R. Verrall, 'Sociobiology: the instincts in our genes', *Spearhead*, March 1979.

38. R. Scruton, 'The case against feminism', the *Observer*, 22 May 1983, p. 27.
39. R. Scruton, *Sexual Desire*, Weidenfeld and Nicolson, 1986.
40. D. Barash, *Sociobiology*, p. 55.
41. E.O. Wilson, *Sociobiology, The New Synthesis*, Harvard U.P, 1975, p. 573.
42. R. Dawkins, *The Selfish Gene*, Oxford University Press, 1976, p. 126.
43. E.O. Wilson, 'Human Decency is Animal'. *Op. cit.*
44. J. Hirschleifer, 'Economics from a biological viewpoint', *Journal of Law and Economics*, 20 (1), 1977, pp. 1–52.
45. D. Barash, *op. cit.* p. 232.
46. E.O. Wilson, *On Human Nature*, *op. cit.* p. 92.
47. R. Verrall, 'Sociobiology: the instincts in our genes'. *Op. cit.*
48. F. Palmer, *Anti-racism*, *op. cit.* The attempt by white parents in Dewsbury in September 1987 to remove their children from a local CofE school now largely populated by 'Asian' children on grounds explicitly stated to be cultural, not racial, 'the right to our own separate culture' is in a similar category. At least this must be seen as an advance over the Nazi Rudolf Hess who claimed that 'National Socialism is nothing but applied biology' – R.F. Lifton *The Nazi Doctors*, Macmillan, 1986, p. 31.
49. C.J. Lumsden and E.O. Wilson, *Genes, Mind and Culture*, Harvard University Press, 1981.
50. F.A. Hayek, *Knowledge, Evolution and Society*, Adam Smith Institute, 1983.
51. For example, Tanner, *On Becoming Human. op. cit.*
52. R. Dawkins, 'Sociobiology: the new storm in a teacup', in S. Rose and L. Appignanesi (eds.), *Science and Beyond*, Blackwell, 1986, pp. 61–78.
53. R. Dawkins, *The Selfish Gene*, and previous reference.
54. For example, S.P.R Rose, R.C. Lewontin and L.F. Kamin, *Not in our Genes*, Penguin, 1984; and P.P.G. Bateson *op. cit.*
55. W.D. Hamilton, 'The genetical theory of social behaviour', *Journal of Theoretical Biology*, 7, 1964, pp. 1–52.

4 DNA and Human Perfectibility

Few topics in biology have generated such controversy as evolution. And when the biological questions have become coupled with those of human society, as has been the case ever since *The Origin of Species* appeared in 1859, the mix has been explosive. Philosophical questions about the 'nature of human nature', religious ones about the distinction between human and other animals, political ones about the biological limits to possible human societies, and ethical ones about the extent to which human evolution could and should be directed by conscious intervention have bubbled up again and again. They took the form of Social Darwinism in the late nineteenth-century, the eugenics movement that peaked in the 1920s and 1930s, and today they appear in forms which vary from the claims of 'human sociobiology' to the dilemmas — ethical and social — raised by *in vitro* fertilisation.

The influence of biological thinking and practice on the social order is clear. But the history of debate in this area cannot be understood as if that influence travelled a one-way street. The whole development of the conceptual and methodological approaches of biologists to human nature, human differences and human origins has itself been profoundly coloured by the social order within which their research programme has been conducted. It is not just that Darwin's understanding of natural selection as the key to evolution was transformed by his reading of Malthus and later became transmuted into Social Darwinism. In an important sense Darwin was himself a Social Darwinist and he approached not merely the human but the non-human living world as such.

It is not my intention in this chapter to explore the theoretical and ideological debates surrounding the propositions listed in the opening paragraphs. Instead, I want to ask a different question, more obviously 'internal' to science itself: what, if any, is the relation between ideas about 'the' genetic material DNA, evolutionary theory, and the limits to human perfectibility? We must begin by looking at DNA, evolution, and development.

The term 'evolution' has been applied both to the process of transformation of species over periods of geological time and to the transition of an individual organism from embryo to adult, a process today called development. Important conceptual issues are at stake in the similarities and dissimilarities between these two processes, the exploration of which lies at the centre of biological endeavour. Understanding the prospects for the perfectibility of humans means coming to terms with both development and evolution.

Genotype and phenotype

The spectacular flowering of molecular biology over the three decades since the discovery of the double-helix structure for DNA revealed how identical copies of DNA molecules could be produced, has made it appear as if DNA was the entire story. But there is a great deal more as well. The key problem lies in the relationship between genotype — the unique ensemble of genes with which each individual starts their life — and phenotype, the properties of that individual as an organism.

The DNA molecule is composed of a series of paired units, called nucleotides, arranged in precise sequence. In the appropriate conditions, and by means of a number of enzymes and other cellular constituents, some parts of this sequence of nucleotides serve as a code which can be translated into sequences of a different type of unit, the amino acids. Sequences of amino acids make up proteins, molecules whose structures and properties help shape all aspects of the phenotype, from the form of individual cells and organs to the complex metabolic processes by which organisms survive, grow and develop. In the early days of molecular biology, in the 1950s and 1960s, most researchers assumed that all of the DNA formed a coding sequence in this way. But it is now clear that such sequences are interspersed with others, some of which have a known function in the control of gene expression; others are without known function and have been variously described as spacer DNA, repetitive DNA, junk DNA, or selfish DNA.

The human genome — the entire complement of DNA present in a human cell — amounts to some 10^9 pairs of nucleotides. This is the genotype, unique to each individual (identical twins apart). It is equivalent to some 10^7 gene-sized bits. However, there are only some 10^5 different human proteins — thus only one per cent of DNA is directly translated into protein. But this is to give a static picture; during development particular proteins are produced only at particular times, others at different times, in an intricately ordered sequence. (For instance at birth and during weaning babies possess an enzyme, lactase, which breaks down the sugar lactose of milk. After weaning, production of this enzyme is reduced and in some [lactase intolerant] populations switched off completely.) In general it is clear that varying proportions of the genome are actively transcribed into protein at any one time and in any one cell. At any time the outward form of an organism, its phenotype, is the resultant of the changing expression of DNA in a particular set of changing environments.

Genes and environment: The lesson of phenylketonuria

It is this interaction of DNA with environment during development — indeed during the whole of an individual's lifetime — which means that there is not a simple, linear sequence by which 'DNA makes RNA makes protein makes phenotype'. It is a tribute to the extreme reductionist theoretical framework within which molecular biology has developed that it was precisely this DNA → RNA → protein sequence, a unilinear and irreversible flow of information, that Francis Crick described as molecular biology's 'central dogma'. Crick's formulation implies a *directional* role to the genes, and a *permissive* and hence essentially non-informational role to the environment. It thus misses the essentially dialectical nature of the interaction of gene and environment during development. First, just as all individuals *except* identical twins have unique genotypes, all individuals *including* identical twins have unique environments. Second, the concept of environment must be understood at many levels. To a gene all other constituents of the cell, including proteins, are part of its environment; since proteins are themselves gene products, this means that each gene of a genotype has all other genes as part of its environment. The cell itself has an environment, which is contributed to by all other cells. In a multicellular organism, like a human, cells influence one another by producing hormones and other signalling substances. And the organism has an environment which includes not merely the physical world, but the biological and social worlds as well. Environment is not the simple reciprocal of genotype.

How do genes and environment interrelate during the development of the organism? From its origins, genetics has had difficulty in dealing with the concept of the organism as an integral unit; instead it is decomposed into a set of more-or-less arbitrarily defined 'characters' which genes and environment are supposed to produce by some apparently simple additive action. (It is hard to escape the conclusion that there is something slightly mystic about the way in which geneticists employ terms like genotype, phenotype, character. They are slippery concepts constantly eluding precise definition, and deriving theoretical weight from genetics' own historical emergence as an autonomous science.) But in fact, genes and environment interact not additively but in more complex ways, expressed in part by the term 'norm of reaction' to describe the range of potential relationships of expression of a genotype to environmental change.

This is why we can only speak of the function or expression of a gene if the environment is also defined. Although geneticists often talk of genes 'for' eye colour, phenylketonuria (PKU), sickle cell anaemia, and so on, this shorthand can be misleading. In the genetic disease PKU, which affects some 1 in every 10 000 children born in the UK, or USA, the affected person cannot metabolise the amino acid phenylalanine, breakdown products of which accumulate in the kidney and urine. The absence of a particular enzyme is normally responsible; in effect, the gene *for* PKU is the *absence* (or alteration) of the gene coding for that enzyme. When geneticists speak of a gene 'for' some phenotypic character, they usually mean the *difference* between the genotypes of people with or without that character. Similarly the famous measure of 'heritability', so misused and misunderstood in the IQ debate, is ostensibly a measure of the contribution of genes to the *differences* between individuals in a given environment.

PKU can also illustrate some other general points about the complexity of relations between genotype and phenotype. The child with PKU, if untreated, will develop with a variety of deficiencies in many body tissues, of which the most striking is mental retardation. The effects of the missing enzyme are different in different body tissues; it all depends on the context whether a particular gene will or will not be important for the development of an organ. Incidentally, despite a good understanding of the genetics and biochemistry of PKU, no one knows why one consequence of the disorder should be mental retardation. The gap between molecular knowledge and knowledge of most body functions is still profound. Sometimes the effect of a gene is experienced only in one organ, as in the gene 'for' eye colour, in other instances its effect is widespread.

But the key to PKU is the phrase 'if untreated'; if the disorder is detected at birth it can be rectified by providing a phenylalanine-free diet. In a phenylalanine-free environment, the gene for PKU no longer produces mental retardation. In other words, the *same gene* in *different environments* produces *different phenotypic effects*. Further the child with PKU is then virtually indistinguishable from other ('normal') children in this respect. Thus *different genes* in *different environments* may produce the *same phenotype*.

Constraints on phenotype

Given this complexity, it is scarcely surprising that we do not yet begin to understand either the rules by which genotypes are translated into phenotypes, or those which describe the effects of changing genotypes on consequent phenotypes, despite a clear knowledge of the mechanisms by which DNA is translated into protein. Furthermore, there is only a very partial knowledge of the constraints on these rules. Despite ignorance of the function (if any, hence the term 'selfish DNA') of 99 per cent of the DNA of the genome, most molecular biologists are convinced that there will come a time when, given a DNA sequence and specified environment, they will be able to predict the phenotypic outcome. Other biologists are skeptical, and point, for example, to the fact that humans and chimpanzees share some 99 per cent of their DNA sequences in common, yet no one would mistake the chimpanzee phenotype for the human. What sort of rules could be embedded in that residual one per cent which could define the difference between the two?

Are there constraints on phenotype which are not 'given' by the DNA? Some, certainly. For instance, there is a limit to the size that an individual cell can grow, determined by the fact that it must be able to communicate with its external environment by way of its surface membrane – food must pass in, waste out. But, as the size of the cell increases, the volume increases as the cube of the radius, while the surface area grows only as the square. The intracellular volume serviced by each square micrometre of surface membrane therefore increases, and there must be a finite limit to this. Evolutionary processes could not therefore produce single-celled organisms as big as a human – something the size of a paramoecium seems about the biggest one can get on this type of structural plan. Multicellularity, therefore, is not given by the DNA, although the DNA to support it has evolved; rather, multicellularity is a structural property of living matter beyond a certain mass. Some developmental biologists (notably Brian Goodwin) suspect that there are many other such 'laws of form' which have to do with phenotypes as varied as the internal membrane structure of cells and the pentadactyl limb of vertebrates, which constrain both development and evolution, albeit yet far from adequately understood.

Fitness

So far, I have been discussing DNA in relationship to one of its two roles: that of development. When Darwin formulated his theory of evolution by natural selection neither the general rules nor the molecular mechanisms of heredity were understood. Darwin's theory was the logical consequence of two irrefutable premises: first, that like begets like (with variations); second, that all living organisms can produce more offspring than survive to reproduce in their turn. As survival or non-survival is not merely chance, those more fit for (or, as we would now say, better adapted to) their environment were more likely to survive to reproduce in turn. Hence more favourable variations would tend to be preserved, and species evolved (changed) as a result of this process of natural selection.

What has DNA to do with this? The modern synthesis of neo-Darwinism (primarily based on the work of Fisher, Haldane and Sewall Wright in the 1930s) argues that it is precise changes in DNA, occurring as a result of random mutations and other internal molecular rearrangements which produce those variations between offspring on which natural selection can act. Just as molecular biology tends to see the fixed sequence of DNA as determining the phenotype during development, so it sees *changes* in the sequence of DNA as altering phenotypes during evolution.

There remain a number of conceptual issues which divide evolutionary biologists and on which passions have run high these past few years. No one doubts the fact that evolution has occurred (*pace* the creationists, who tend to pretend otherwise). The debates are over the mechanisms of such change.

Expressed in the bald form offered above, natural selection theory seems obvious, incontrovertible (which is why it has been accused by some of being tautologous and hence not a 'proper' scientific theory). Yet the problem remains that although it provides a good explanation of how species get better at doing what they do, it provides a poor and stumbling explanation of the formation of new species. Hence the arguments about the need for different mechanisms to account for the mode and rate of formation of new species. Other issues such as why the phenotypes of some species, such as some molluscs, seem to have remained unchanged over hundreds of millions of years despite genetic changes, have helped lead Stephen Gould and Niles Eldredge towards their version of evolution theory, punctuated equilibrium.[1] But my major concern here is with the concept of adaptation or fitness, and whether it is appropriate to argue that all selection ultimately occurs at the level of the DNA.

The concept of fitness cannot be absolute; it must depend on the environment. For instance, in modern Western societies, the gene which is responsible for sickle cell haemoglobin is a deleterious mutation; heterozygotes (those who inherit one copy of the 'sickle' gene, and one of the 'normal' gene) are at some disadvantage, and homozygotes (those who inherit two copies of the 'sickle' gene, one from each parent) are likely to have a short life. So why has the trait not been eliminated during human evolution? The answer generally given lies in the fact that to be heterozygous for sickle cell gives some immunity from malaria, hence in malarial regions to be heterozygous for sickle cell is an advantage.

As Darwin's ideas became accepted during the first part of this century, concern began to grow that health care and social amelioration were preserving certain human traits which would otherwise have been eliminated — for instance, a genetic propensity to shortsightedness. Eugenists began to worry about genetic deterioration, the prospect of an increasing 'genetic load' of deleterious genes.[2] Such an argument ignores the fact that, because the human genotype produces persons who exist in social organisations, the modern human environment includes spectacles. The consequence of having a genotype which can participate in the creation of a society which can produce spectacles is to change the environment such that genes 'for' shortsightedness are no longer less fit than genes 'for' 20/20 vision. There are both social and genetic consequences of such changes. The increasing population of diabetics, which occurs partly as a result of the fact that certain genotypes are more prone to diabetes, so that their preservation by insulin therapy may increase the number of potentially diabetic offspring they produce, is but one example (these are environmental factors such as diet at play in diabetes too, but that is a separate matter). But to say there are consequences does no more than to state the obvious; it does not necessarily imply loss of fitness, in the sense of ability to leave offspring in an environment which includes the technology to manufacture and distribute insulin.

This points to a further complexity in the use of the term 'environment'.

Developmental biology has tended to see the environment as a fixed background against which the internal genetic programme of the organism unrolls. Evolutionary biology has tended to see organisms as passive recipients of an environment which presents them with challenges which they either pass or fail, depending on their genes. Both metaphors ignore the active part which any organism plays in changing and transforming its own environment. As Richard Lewontin, Leo Kamin have previously pointed out[3], if one puts a ciliated bacterium into a glass of water and adds a drop of glucose the bacterium will swim towards the glucose, seeking a glucose-rich, in preference to a glucose-poor, environment. It will metabolise the glucose, taking some into itself, and excrete waste products which will change the environment. One such product may be acidic and the bacterium will then move away from the now acid environment. All organisms, even the simplest, actively seek, interpenetrate with and transform their environment. Humans do so consciously. This is one of the senses in which we can say that we make our own history, though in circumstances not of our own choosing. And it is one of the reasons why a knowledge of the evolutionary or historical past does not allow us to predict the course of the evolutionary or historical future.

A further problem with the idea of fitness and adaptation is to decide which aspect of a phenotype is the character on which selection might be operating; the past few years have seen a host of evolutionary fables being produced to 'account for' particular presumed phenotypes, from the shape of the human chin to children's alleged dislike of spinach. To give an example, haemoglobin is characterised by its red colour. Is its redness a phenotype which is adaptive and has been selected for? A story about how red acted as a warning signal and hence registered danger, alerting individuals close to a wounded victim of the need to escape, could be constructed. Yet it is much more likely that the red colour of haemoglobin is a contingent property, a consequence of the fact that iron-containing compounds are red. What has been selected for in haemoglobin is a molecule with very high oxygen-carrying capacities.

Programmed by our genes?

Some molecular biologists, and following them, sociobiologists have argued that because DNA is the genetic material and because only DNA molecules possess the property of self-replication, then all selection acts ultimately at the level of the DNA. It is this which has led to the extravagant claims that an organism is 'merely' DNA's way of creating more DNA or that humans are 'lumbering robots programmed by our genes' to quote the phrase used by Richard Dawkins, in *The Selfish Gene*. In reality however selection must act at a multitude of levels. Individual gene-sized lengths of DNA may or may not be selected in their own right, but that DNA is expressed against the background of the entire genotype; particular assemblies of genes or whole genotypes must therefore themselves represent another level of selection. Further, the genotype exists within a phenotype, and whether that phenotype survives or not depends on its interaction with others. Hence it will only be selected for against the background of the population in which it is embedded. And if a 'fitter' species evolves, an entire population may be eliminated irrespective of the selective advantage of particular genes in the genotype of any individual member of that species compared with other such members.[4]

These points are important because much attention has been addressed to the question of the evolution and genetic determination not merely of individual physiological traits but of complex social behaviours. For sociobiology such behaviours as mate selection are to be regarded as phenotypes under genetic control. But why do behaviours which are apparently disadvantageous to the organism evolve? One bird in a flock may give a warning cry of the approach of a predator, drawing attention to itself but potentially saving its fellows. How can such apparent altruism be a product of 'selfish DNA'? The answer popularised today by sociobiology is that by potentially sacrificing itself, the bird may preserve its close relatives, which share some of its genes, hence such behaviour may be in the 'interests of the genes' if not of the individual. A theory of the genetics of kin-selection, developed by Hamilton has grown up around such behaviours, but despite a host of more-or-less vulgar popularisations, its relevance to the human situation (and indeed to most animal behaviour) is at best obscure. (It must be emphasised that while much of the wrath of the critics has been concentrated on the crude reductionism of human sociobiology and its links with the ruling ideas of a class, race and gender divided society, sociobiology's account of non-human behaviour is virtually as crudely reductionist, as Philip Kitcher, amongst others, has shown in his book *Vaulting Ambition.*[5]

Improving humans

Let me try to pull these threads together in the context of human evolution. It is sometimes argued that, with the arrival of *Homo sapiens*, biological evolution is no longer important but has been replaced by 'social evolution'. Such terms have been used by political and social theorists of the right, from Herbert Spencer to Friedrich Von Hayek and, of the liberal left, like Julian Huxley. However, the analogy is unhelpful; evolution in the biological sense is an inevitable consequence of being alive, irrespective of disputes over details of its mechanism. But the processes of change in human societies are not comparable with the process of natural selection; they do not involve self-replication of molecules or organisms. Social change is the product of the interaction of biological, economic and cultural forces and conscious action, which need to be studied in their own terms, not by spurious analogy. It has also been fashionable to argue that human societies are the inevitable products of the properties of self-replicating genes; that selfish genes make selfish people and hence that although we may not live in the best of all conceivable societies, we live in the best of all possible ones. This so-called 'panglossian paradigm' has been the stuff of the vulgarized biology of the ideologists of the New Right discussed in the previous chapter.

Such claims are as fallacious as were those of an earlier generation of eugenists and political theorists of the 1930s.[6] We do not know and cannot predict the limits to human nature set by the human genome. The only thing we know is that there *are* no obvious limits. The sorts of societies we build and the ways in which we view the world are of course shaped by our biological natures. The human genome ensures that we live for a maximum of a hundred years or so, are bipedal, have language, and are around 1.5–2 m in height; that we can see the world through eyes sensitive only to radiation of a small range of wavelengths; that we cannot sprout wings and fly (although some of these limits, such as that which stops us becoming angels, are set by structural considerations rather than DNA). If we were only a few centimetres in height, or could see in the infra-red or ultraviolet or fly like a bird, we would perceive the world differently and build different societies. But the extraordinary thing about the human genome is that it permits us to build instruments which enable us to sense wavelengths that the eyes cannot see and build machines that allow each and every one of us to fly without becoming an angel. Such societies, such machines, do not go against human nature, for there can be no such antithesis. It is the human genome which makes possible the brain, language, social and tool-using ability which enables us to create our own history. It is our genome, therefore, which allows us — which insists almost — that we constantly transcend the limits apparently set by that very genome, enabling us to continuously reconstruct our future on the basis of our past, to have the freedom to make our own history.

The future

The new genetic knowledge, product though it be of a science conducted in the name of profit and of theories refracted by ideology, is part of the process of making that history. This is not to embrace a naïve progressivism. Still less is it to accept that a science which is not controlled by, not in the hands of, the people can automatically be beneficial. Knowledge in the hands of the dominant class, race and gender is knowledge used in support of that domination. Nonetheless, one is entitled to ask what scope that knowledge offers for changing the future pattern of human society or human evolution?

To the former, my answer must be 'not much'. The social relations of human societies, if not their knowledge, are likely to be changed more by social, economic and political actions than by knowledge of DNA, for just the same sort of reason that Florence Nightingale did not require a knowledge of germ theory to propose changes in hospital design which would reduce the mortality of hospital patients, or early farmers a knowledge of genetics in order to improve their wild crops. As to human evolution, the air is certainly thick with promises — or threats.[7] Both positive and negative eugenics are again in discussion. On the one hand, sperm banks to preserve the offerings of males perceived as especially gifted, such as what was to have been called the 'Herman J. Muller genetic repository' in California until his widow protested. On the other hand, the offer in the first instance of amniocentesis, and later possibly gene therapy for the replacement of particular disordered genes in the foetus.

Such possibilities have raised a wide debate. For example, the Warnock Report, commissioned by government in the UK, summarised both the scientific issues and the ethical questions surrounding one part of the terrain of genetic engineering — *in vitro* fertilisation and experimentation on human embryos.[8] A wide debate, centred on the morality of IVF-related techniques has followed.[9] In my opinion the wrong question is being asked, because it endeavours to turn what are issues of priority in health care into those of abstract ethics. Instead, one should ask the prior question, which the *in vitro* fertilisation techniques are presumably designed to help answer: how can we increase the number of wanted, healthy babies? What prevents wanted, healthy babies surviving? In Britain the perinatal mortality rate (that is, the number of babies dying at or just after birth) is much higher in certain geographical areas such as Liverpool, than others, such as Hampstead. There is a very much greater chance of a baby not surviving if it is born to a mother in poverty, or in the manual working class than if it is born to a wealthy or upper-middle class mother.[10] The conclusion must be that if we want to save babies, we can do so best by applying known social, economic and health care improvements to deprived geographical areas and classes in Britain. *In vitro* fertilisation is a method which is, willy-nilly, only of relevance to a small number of relatively privileged mothers. The language of priorities says that we shouldn't get excited about that new set of techniques until we have addressed the question of how we save babies we *know* statistically will die from lack of application of quite simple preventative and health care measures.

Ignoring for the moment the high-tech glamour which surrounds such medical advances aimed at individuals, as opposed to collective measures directed towards communities; as a way of circumventing infertility the techniques of IVF are here to stay, though their contribution to human evolution is likely to be marginal. And, despite science fiction speculations about the prospect, the cloning of humans is unlikely to prove technically possible in the near future. Fascinating philosophical speculations about the implications of such procedures will doubtless continue, but our 'brave new world' is unlikely to be that envisaged by Aldous Huxley.

In fact the limits to the prospects of gene manipulation are likely to be set by theory more than techniques. They were foreseen as long ago as the 1930s by Muller, Haldane and other geneticists of the period. First, most phenotypic conditions of medical or social interest are the products of multiple gene interactions, differentially expressed depending on the environment. It will be easy to predict the hair or eye colour of the offspring sired by a Nobel prizewinner's deep-frozen sperm; we have no idea about which of many different multiple sets of genes, in which of an almost infinitely varying set of environments, might be associated with intelligence, even assuming that the term 'intelligence' represents a measurable phenotype. Put crudely, we do not know what to breed for, but we can make a fairly shrewd guess as to the sorts of environment to avoid so as to improve the chances of a child with any old genotype becoming a potential Nobel laureate — should that be our goal.

On the side of negative eugenics, there are several single gene disorders, where amniocentesis and genetic counselling can provide advice today and perhaps gene therapy tomorrow. Huntington's chorea, Duchenne muscular dystrophy and some blood diseases are often quoted. But it should be emphasised again that for the overwhelming majority of diseases and distresses which humans suffer there are either no specific genetic components (by which is meant that no particular genotype predisposes a person to the disease more than any other genotype) or a multiple and diffuse genetic contribution, beyond the useful scope of even speculative therapy, even assuming that the long chain of mediations between genotype and phenotype discussed above were more fully understood. Molecular biologists and geneticists tend, not unreasonably, to concentrate on the striking genetic cases such as Huntington's chorea. But these must be set in context. For example, only perhaps some five per cent of the 16 000 or so new admissions for mental handicap or mental retardation each year in Britain have a clear genetic association. At the same time in some diseases once believed to have a clear genetic predisposition, like schizophrenia, the evidence is once more under serious question. (See chapter 8.)

Nonetheless, the new technologies do raise questions which go beyond issues of priority in resource allocation. As in all areas of our social existence, they confront us with questions of control and of power. Where the majority of research is done in the interest of production and profit on the one hand, and social control on the other, and where that research is largely done by or under the control of a privileged class, gender and race, we have no guarantee, to put it no more strongly, that it is being done and used in the interests of the people as a whole. If we reject the technological imperative which argues that we are driven wherever a neutral science leads us, and equally transcend the ideological mystifications which surround much of modern-day genetics, we can begin to confront directly the problem of creating a science for and by the people.

Notes and References

1. S. J. Gould, 'Darwinism and the expansion of evolutionary theory', *Science*, 216, 1982, pp. 380–7.
2. This concept was introduced by the communist geneticist Hermann Muller in the 1930s. See for example his book *Out of the Night*, Vanguard Press, 1935.
3. S.P.R Rose, R.C. Lewontin and L.F. Kamin, *Not in our Genes*, Penguin, 1984.
4. See Patrick Bateson's debate with Richard Dawkins in *Science and Beyond*, S. Rose and L. Appignanesi (eds.), Blackwell, 1986. pp. 79–99.
5. P. Kitcher, *Vaulting Ambition*, MIT Press, 1985.
6. See for example A. Chase, *The Legacy of Malthus*, University of Illinois Press, 1980; D.D. Pickens, *Eugenics and the Progressives*, Vanderbilt University Press, 1973; B. Evans and B. Waites, *IQ and Mental Testing*, Macmillan, 1981.
7. J. Glover, *What Sort of People Should There Be?*, Penguin, 1984; Z. Harsanyi and R. Hutton, *Genetic Prophecy*, Granada, 1982.
8. The Warnock Report, HMSO, 1984.
9. See for example, H. Rose, 'Victorian Values in the Test-tube', in Michelle Stanworth (ed.) *Reproductive Technologies: Gender, Motherhood and Medicine*, Polity, 1987.
10. P. Townsend and N. Davidson, *Inequalities in Health*, Penguin, 1982.

5 Genetic Engineering: The new arms race

Media attention in recent years has been focussed on Star Wars, the Zero Option and the complex diplomatic dance of the disarmament negotiators around intermediate range missiles; but there is another increasingly sinister arms race under way. In 1972 the Biological and Toxin Weapons Treaty was hailed as a model agreement, banning permanently one whole class of weapons of mass destruction. When that treaty was signed, the revolutionary new techniques of genetic engineering were scarcely a science-fiction glimmer in molecular biologists' eyes. In 1987, the United States Department of Defense is directly funding more than $150 millions-worth of research annually on potential new chemical and biological weapons systems and the defences against them, and is involved in a bitter battle against the environmentalist lobby for approval to build a $300 million testing facility at Dugway, in Utah. So serious is the new threat that it dominated the discussions at the 1986 conference in Geneva held to review and hopefully strengthen the 1972 Treaty. Meanwhile, in May of that year, the NATO Council of Ministers, Britain and West Germany to the fore, finally gave the Reagan administration a prize it has been seeking since 1980, when it adopted chemical weapons as a NATO 'force goal'. This opens the way to the first phase of an American production programme for a new generation of chemical weapons, primarily binary nerve gas agents, estimated at some $15–26 billion over the next decade. Thus an entire new area of biological research is becoming militarised just as has happened with computer and information technology with Britain's Alvey and Reagan's Star Wars programmes. How has this new arms race come about, and what are its implications?

To understand the implications of the new weapons they need to be seen in the context of how they have been developed and used in the past. A million gas casualties, from chlorine, phosgene and mustard, in the First World War paved the way to the signing in 1925 of the Geneva Protocol banning the first use of chemical and bacteriological weapons. But the Protocol didn't prevent research, development and stockpiling, and although some forty countries signed it, some were very slow to ratify it — the USA took nearly fifty years!

The first generation of gases caused choking, vomiting, burning of skin and eyes; high concentrations were needed and you could get a lot of protection from a gasmask. But in 1936 Gerhard Schrader of IG Farben discovered a new class of agents, the nerve gases, which can be absorbed through the skin and are 100–1000 times more toxic than chlorine. 250 000 tons of three types of the gas were produced in bulk in Germany but never used. After the war the gases and the technology were shipped back to the USA, USSR and the UK.

British and American research at this time concentrated on biological weapons (BW), either *toxins* — the poisonous products of living organisms such as bacteria, fungi, plants, snakes and fish — or the micro-organisms themselves, from smallpox to cholera and plague. British research favoured anthrax. Gruinard island, off the NW Scottish coast, was infected experimentally in 1944 and remained contaminated until, early in 1987, the Ministry of Defence embarked on a clean-up programme. Churchill is said to have pressed for the development and use of anthrax bombs.

The Japanese went much further. They carried out direct experiments on Chinese and American prisoners of war, infecting them with anthrax, plague, cholera and others, as well as some small-scale field use in China. After 1945 the USA hushed up the Japanese experiments, blocked a Russian request for war crimes trials and shipped the researchers back to the American chemical weapons (CW) base, Fort Detrick, in Maryland. A new wave of research began. In the 1950s Porton, Britain's CB research station in Wiltshire, picked up on some work begun by Shell and came up with the first of a new generation of nerve gases, the V-agents, and a new tear gas (CS). The British also pioneered the use of plant-destroying substances or *defoliants* in their Malayan campaign of the late 1940s and early 1950s. But through the 1950s CB weapons were a poor relation to their nuclear counterparts.

Then came the Indochina war with the massive use by the US of 'non-lethal' CS and the defoliants. The anti-war movement created the climate in which President Nixon agreed to the new treaty banning research and development of all biological and toxin weapons (but *not* chemicals) in 1972. BW research was said to have ceased at Dugway, and at Porton the Microbiological Research Establishment was demilitarised and now forms the base for public health work and the biotechnology complex Porton International. The Treaty should have been followed by a CW treaty as well, but to no one's great surprise it was not. (This is why Edgewood Arsenal in the USA and Porton's *chemical* warfare research centre remain military establishments.) The point is that biological weapons, unpredictable and slow in their effects, are of much less military utility than are chemicals. The BW treaty could be welcomed by the military, whereas there were more powerful interests at stake (including the chemical industry) when it came to the banning of the development or stockpiling of chemicals.

Well-authenticated cases of the use of CB weapons have been rare since the end of the Indochina war. The Portuguese (or perhaps the South Africans) used defoliants in Angola; the Iraquis have used mustard gas against the Iranians and are now said to have nerve gas production facilities available. Unverified are Cuban claims that an outbreak of swine fever in the early 1980s was deliberately started by CIA action, and Eritrean reports that the Ethiopians possess Soviet-produced nerve gas and have launched several CW attacks.

Then there has been the great 'yellow rain' saga. The USA still maintains that deaths among dissident tribes in Laos and Kampuchea (and with less certainty in Afghanistan) have occurred as a result of attack by the Vietnamese with a fungal toxin (so-called 'yellow rain') provided by the Russians. But Harvard CW expert Matthew Meselson was able to confirm Australian reports that yellow rain samples are largely pollen; the mysterious spray is in fact the product of mass 'cleansing flights' by defaecating bees. The American State Department has begun to back-pedal, at least about Afghanistan, but its official briefing document, issued to British MPs in 1985 and repeated at the 1986 Geneva discussions, still repeats a version of the yellow rain story.

If the yellow rain allegations *had* been true, they would of course represent a clear breach of the 1972 Treaty, and this is exactly what the USA claims. It goes further, too in arguing that the USSR is actively engaged in preparations for CB warfare. The Soviet Union is said to have large stockpiles of chemical weapons – according to American estimates, anything between 30 000 and 300 000 tons. Such estimates are unconfirmed, and the high figures are discounted by leading CW experts such as Sussex's Julian Perry Robinson, who believes the figures are likely to be more on a par with America's own stocks of 42 000 tons, held partly in West Germany. Soviet troops *are* routinely trained in the use of CB protective clothing, but then so are NATO's.

Then there was the Sverdlovsk incident. In 1979 there was an outbreak of anthrax in Sverdlovsk which the USA maintains was the result of an explosion at a BW plant ('Military Compound 19'). The Russians insist that the deaths were due to the sale of contaminated meat on the black market. Anthrax *is* a potential BW agent, but outbreaks do occur naturally. Both Meselson and Zhores Medvedev, a well-known critic of Soviet military and civil science, accept the Soviet version, and at the September Treaty Review Conference, the Soviet delegates surprised their Western counterparts by offering the opportunity to put questions directly on the alleged incident to Nikolai Antonov of the Ministry of Public Health in Moscow.

Early in 1984 an extraordinary series of articles appeared in the *Wall Street Journal*, previously the medium for the most extravagant of the yellow rain and Sverdlovsk stories. These articles claimed that there is a major new threat of the militarisation of biotechnology in the Soviet Union. They referred to the doyen of Soviet molecular biology, Yuri Ovchinnikov, as playing a leading role in this development, an idea laughed to scorn by Medvedev. The publication of these articles was followed by a cover story in *Environmental Action* magazine pointing out the implications of such publicity on plans for increased spending on CB weapons research by the American military. The article refers to the strange appearance of a long script sent to a number of Western scientists in 1983 by a Russian mathematical epidemiologist, Leonid Rvachev, attempting to model transmission of infection on a world-wide scale. It claims success in modelling the pandemic spread of 'Hong Kong' influenza in 1968.

At first sight this is no more than an elaborate and not very sinister model-building exercise. But some of Rvachev's correspondence apparently referred to the implications of such models for biological warfare, and suggested that the modelling could help monitor the spread of potential biological weapons. The hyperbolic Jeremy Rifkin, of the Foundation on Economic Trends in the USA, active in the campaign against all forms of genetic engineering research, seized on the Rvachev model as making possible 'a guided missile system which turns biological weapons into precisely targetable agents'.

It is this mix of rumour, extrapolation and assertion which has formed the background to the US claim that there is a danger of a 'CB weapons and research gap' opening up with the Soviet Union. (Despite the fact that present American stocks of 20 000 tons of nerve gas are enough, theoretically, to kill the world's population a thousand times over.) It has led to a security clamp-down on discussion of new developments in molecular biology by American researchers at meetings where foreigners are present, and a powerful lobby for the expansion of CB research in the USA. Assiduously propagated by American Defense and State Department officials, the claims have been used to propel other NATO countries into acceptance of the need for the new generation of binary nerve gases and the resumption of weapons stockpiling and research.

The Geneva Protocol does not inhibit research on new chemical weapons. The 1972 Treaty bans research on biological and toxin weapons intended for offensive purposes, and on stockpiling, although it does permit small-scale production of offensive agents for defensive studies. The trouble is that the distinction between offensive and defensive research is somewhat metaphysical; almost any type of work can be justified under the rubric of investigating agents which an enemy may be thinking of using against oneself. As we will see, it is on this basis that new research in the USA, in large measure based on the new genetic engineering methods, has indeed been commissioned. Further, it is unfortunately also true that the nature of biological weapons is such that it is not necessary to have large stockpiles prepared in advance; provided a delivery system has been tested, containment facilities are available, and the production route for an organism is known, then it would be possible to produce militarily significant quantities from tiny amounts of type culture within a very short space of time. It *is* possible to imagine genuinely defensive BW research, for instance on new forms of civilian protection or shelters, but this seems to be very low on the list of present Pentagon priorities.

Whatever the demerits of the American system and the power of the Pentagon, it has one particular strength lacking either East of the Elbe or for that matter in Britain or France — it is open and public information is much more readily available than elsewhere. This means, for instance, that the Department of Defense publishes lists of its research grants and contracts, which make it possible to at least catch a glimmer of what it perceives as military priorities in biotechnology. A list of some sixty such contracts has been published in the American magazine *Genewatch* in 1985[1] from which it begins to be possible to construct a taxonomy of prospective new and improved weapons (the number has crept up steadily since). By contrast the British Ministry of Defence refuses to provide such lists, saying only that in total it has some sixty-five contracts let via Porton or relevant to chemical and biological weapons at British universities.

Nonetheless, analysis of the published lists of research contracts by British universities does enable some of the contracts to be identified.[2] Thus apart from many projects on organic syntheses, the MoD has contracts out for research on opiate receptors, concerned with brain regulation of pain, at the University of Bath, while Surrey biochemists are studying the interactions of acetylcholinesterase poisoning (that is, nerve gas) with opiates. There is work on cell counting methods at Cardiff, lung and other toxin research at University of Wales College of Medicine, and studies of skin absorption at the University of Wales Institute of Science and Technology. Anthrax and brucella antibodies are being researched at Glasgow, and early events in viral infection at Birmingham.

While many of these projects may relate to the extension of what we might call 'conventional' CB weapons, from the list of American contracts it is possible to identify three broad areas of new research. The first lies in developing the existing range of chemical weapons. Biotechnological methods are not going to supersede chemical ones for the production of nerve gases or other agents, but the site of action of the agents is of course biological and involves enzyme systems. For instance nerve gases act on the acetylcholine neurotransmitter system, and there has been military interest for many years in the workings of this enzyme and related aspects of the transmitter-receptor systems. The nerve gases are organophosphorus chemicals, and a quick literature scan reveals a steady production of Porton- and Edgewood-originated papers on organophosphorus-acetylcholinesterase interactions.

Such studies can lead in two directions of course. One would be the development of still more effective or even perhaps selective anti-cholinergic substances. Alternatively, the concern could be with possible antidotes. The standard treatment for nerve gas exposure is self-injection with atropine, a hazardous procedure because this substance is itself toxic. So alternative detoxifying agents would be useful. It comes as little surprise, therefore, to find that the American DoD has placed a number of contracts for the isolation of the acetylcholine gene, its insertion into that standard genetic engineer's workhorse, the gut bug *E. Coli*, and subsequent large-scale production of the enzyme. The contracts for work along these lines in the USA include one at the Salk Institute and others with private companies. In addition there are contracts with Drs Soreq and Silman at the Weizman Institute in Israel and Dr Harris at Inveresk Research in Edinburgh. A related set of contracts refer to the possible development of catalytic enzymes for the degradation of organophosphorus compounds.

The second possible area of CB interest in which genetic engineering techniques could be envisaged is in the large-scale production of existing toxins, or the engineering of new ones. Naturally occurring toxins have been prime candidates as agents both for battlefield use and for sabotage and assassination for many years. In 1978 the Bulgarian dissident exile Georgi Markov was killed in London, stabbed with an umbrella tipped with the toxin ricin. Ricin, from castor beans, is almost as toxic as one of the V-agents; botulinus and tetanus toxins considerably more so. Toxins are biological products (often but not always protein in nature) rather than biological weapons, which is why the 1972 Treaty classed them in a category by themselves. Whether genetically engineered toxins are covered under the Treaty, or whether they are merely a special type of chemical weapon, is a nice legal quibble which need not concern us here.

One could envisage genetically engineering such molecules in order to enhance their toxicity or alter their stability, although it is hard to see the point of doing so. Botulinus and tetanus toxins are products of micro-organisms and therefore readily available in bulk, but others are much rarer, such as saxitoxin, or tetrodotoxin, derived from fish and snake venoms respectively and whose action is, like the nerve gases, directed at neurotransmitter systems. If their production were considered desirable, then recombinant technology could be used to insert the genes for their synthesis into *E. Coli* so as to generate them in bulk. Then of course there is the question of antidotes or vaccines. The open literature published since 1983, mainly from military-sponsored researchers, includes work on cloning and expressing genes for anthrax and its expression in *E. Coli*, the sequencing of botulinum type A toxin, and the cloning and sequencing of cholera, shigella and diphtheria toxin genes.

Of the list of Department of Defense contracts five relate to toxins, including anthrax, unspecified snake venoms and *E. Coli* exterotoxins. The contracts, at the US Army Medical Research Institute of Infectious Diseases, the Walter Reed Hospital and Brigham Young University, all include the key phrase 'for the production of vaccines'. The circumstances under which snake bites would be a serious hazard for American troops are not clear to me.

We move on now to the third category, work directly on infectious or potentially infectious organisms themselves. Once again, there is already a vast range of agents which have been studied for potential BW use. The criteria for a good agent include consistency of effect in producing death or disease at low concentration. It should be highly contagious, producing its effect quickly, and the population under attack should have little immunity to the agent and no ready access to prophylaxis. Above all, a potential agent should be one against which a user can readily protect their own side, by vaccination or similar means. Whereas until the new technology was available most agents which met these criteria were bacteria or fungi, the new techniques have also triggered an interest in viruses, if only because their genetic material can now be studied within host bacteria.

It is a fair guess that the BW research establishments on both sides have studied vaccine production against a variety of potential existing BW agents, so there would be some mileage in attempting to tailor organisms so as to alter their antigenicity and thus nullify any potential vaccine. The existence of viruses with highly variable antigenic structures — such as the AIDS agent HTLV-III — could prove helpful in this regard. (There are persistent claims that the AIDS virus was itself the accidental result of a genetic engineering or even BW experiment that went awry, but there are abundant reasons for discounting any such suggestion.)

Other potential manipulations of biological agents could include genetic tailoring to increase pathogenicity of viral virulence, or even to take hitherto non-pathogenic organisms and transfer pathogenicity to them — a sort of public-health-in-reverse version of the fear of the accidental consequences of genetic engineering which led to the original moratorium on research and the development of containment facilities in the mid-1970s. The markers used for the diagnosis of existing pathogenic organisms could be modified, or drug-resistance artificially induced. Finally, genetic engineering could be used to make organisms easier to produce or store.

A further area of interest lies not in organisms which affect humans, but those which affect animals or plants, hence potentially having a crippling effect on a country's agriculture and economy without necessarily being directly engaged in overt conflict or indeed without raising suspicions of deliberate sabotage. This idea lay behind both American statements that Mid-West monoculture was vulnerable to KGB sabotage and the Cuban claims that their recent outbreaks of swine fever had been deliberately spread. New or modified BW agents would be relevant here too.

Again, the DoD contracts provide indications that the American military is thinking along these lines. Granted that nearly all refer to the purpose of the work as for vaccine production or related protective measures, the contracts discuss leishmanias, trypanosomes, salmonella, shigella, gonorrhea, typhus, Rift valley fever, dysentery, dengue fever, malaria, encephalitis, lymphocytic choriomeningitis, hepatitis, arboviruses and pox among other less-familiar disease-bringers. Genes for surface glycoproteins (which are among other things responsible for antigenicity) of Rift valley fever virus and of Punta Toro virus — one of the agents responsible for phlebotomus fever — have been cloned and sequenced at the US Army Medical Research Institute of Infectious Diseases.

This accounts for virtually all the published DoD contracts, except for a group which refer to marine fouling projects, and others that speak of cloning the eye's visual pigments, including rhodopsin. One of the rhodopsin contracts is let to MIT's Nobel prizewinning molecular biologist H.G. Khorana. It is difficult to think of an obvious military use for such studies, though another US Nobelist for work on visual pigments, George Wald, says that he was approached many years ago by DoD to consult on a project to develop blinding agents. On the other hand people interested in biocomputing systems have been considering the possible use of rhodopsin as a biosensor molecule — that is, one which can switch its state in response to incoming energy pulses — so maybe its relevance lies in quite another direction.

Examining research contracts like this produces a sort of phrenological approach to the range of military interests in biotechnology and genetic engineering in the USA. Even if one discounts the *Wall Street Journal* reports, it would seem to me unlikely that prudent Soviet military researchers, knowing of these developments, could avoid developing a parallel programme, although the known Soviet weakness in molecular biology and genetics might limit its scale.

Such considerations certainly underlay the discussions at the Treaty Revision Conference in Geneva in September 1986, which agreed a series of 'confidence-building' measures. These included proposals to exchange information on the activities of research labs with containment facilities for high-risk genetic engineering research, and on the outbreak of unusual toxin-related diseases, and on the exchange of scientists engaged in such research.

Meanwhile, the fate of the proposed aerosol facility at Dugway, which is intended to provide P4 containment facilities for testing sprays of biological agents, hangs in the balance. The proposal to construct the facility was slipped through the relevant congressional committee in 1984 and its implications only subsequently realised. Energetic legal action spearheaded by Jeremy Rifkin halted the project in its tracks, whilst a full environmental impact analysis is prepared and debated.

The hazard of new escalation of the East-West arms race, and of the increased militarisation of biological research, is clearly acute. What is puzzling in this situation is to discover where the driving force for these new developments lies, because a cool look at the actual potential of the new weaponry makes it far from obvious.

Even as far as the new generation of nerve gases are concerned, American military scenarios for their use, and the destined home for the binaries, is Europe. Yet in practice, since 1918 chemical weapons have never been used in war in the type of East-West conflict possible in Europe. The characteristic use of the agents has been by technologically advanced powers against ill-equipped peasant or Third World guerilla forces. The Italians in Ethiopia, the Japanese in China, the British in Malaya, the USA in Indochina, even the Iraquis in the Gulf War, are all in this category. Indeed if the yellow rain saga *were* true it would merely make the point even more strongly. CB agents are of course weapons of mass destruction, despite the assumption that they would be battlefield weapons in Europe. It is far from clear what adding nerve gases to the nuclear arsenals of Europe would achieve in military terms. The argument is apparently that it would force Warsaw pact troops to fight wearing full CB protective gear, but as they are already committed to nuclear protection it doesn't seem to me (as no sort of military specialist) that it adds very much.

But whatever the actual or potential use of chemical weapons, they can at least be said to make *some* sort of military sense. Biological weapons never have done. They are by their nature uncertain in their effects, slow to act, dependent on natural means of propagation like wind, water, and later contagion. Once released, their effects cannot be limited to a war zone but may spread uncontrollably. Protection of one's own side, especially the civilian population, is difficult if not impossible, demanding complex and costly mass vaccination and/or shelter programmes.

This is probably why military strategists have always rejected their use, despite occasional experiments like the British settlers' gift of smallpox to the north American Indians or the Japanese use of anthrax. Churchill may have wanted an anthrax bomb, but not even the advocates of saturation bombing, the men who called the firebombs down on Dresden and Tokyo and unleashed atomic explosions at Hiroshima and Nagasaki, went along with him.

So why is the CB genie emerging from its bottle once more? The pressure for the binary nerve gas programme may be partly military, but it is known also that segments of the American chemical industry, with shortened order books, have lobbied for it, despite the fact that the influential US General Accounting Office has criticised the weapons as impracticable and unreliable. So far as the biologicals and toxins are concerned, I believe that, paradoxically, some of the blame must lie with molecular biologists themselves. They — we — became so excited at the potential powers and hazards of genetic engineering in the early 1970s that we laid claims to having developed a technology of great power. The biotechnology hype that followed turned university molecular biologists into millionaire shareholders in the new biotech companies, even though they have not yet produced much in the way of viable products as opposed to paper fortunes. It immeasurably strengthened the university-industrial complex, bringing secrecy, patents and chequebook-led research into once open labs. And inevitably it interested the military. What is pushing BW research forward again is an unholy alliance of cold-war and pro-rearmament political rhetoric, military visionaries, and an unscrupulous scientific entrepreneurialism unable to resist the combination of power and grant money that the military boondoggle offers.

Even if I am right in locating the source of the new arms race, it in no way lessens its significance. It is true that, just as elsewhere in the field of biotechnology there are gaps between promise and achievement (the difficulties that a major pharmaceutical company like Eli Lilly has faced in successfully bringing genetically engineered human insulin into market production is but one example of the problems that may be faced if significant investment in the new BW weapons systems were really to begin). But the argument of a 'scientific and technological imperative' is powerful. It was after all J. Robert Oppenheimer, and *not* Edward Teller, who when shown a practicable method for producing the first H-bomb, said that 'it was technically sweet; we had to go ahead and do it'.

To prevent this new arms race, molecular biologists could learn from their colleagues among the physicists and computer engineers who have made public their refusal to work on Star Wars. But self-declared moratoria are not enough without at the same time a widespread popular demand to the Geneva negotiators to achieve the negotiated strengthening of both chemical and biological warfare treaties which is now urgent. In the meantime unequivocal statements from European governments to the effect that they will neither support research directed towards new CB weapons, nor permit their stockpiling, would help restrain the rearmers.

Notes and References

1. The full list of US contracts appears in *Genewatch*, 2 (2), August, 1985.
2. I am grateful to Rob Evans, of CAMROC, the Campaign Against Military Research On Campus, for lists of the British MoD contracts.

6 Animal Rights and Wrongs

We used to live next door to an elderly pacifist. She would invite our young sons in to help in important tasks like stuffing bread rolls with aniseed to distract the dogs before going down to demonstrate outside Porton; meantime, she would quiz them about whether it didn't worry them to have a father who experimented on animals as part of his job? If his younger brother were ill, replied my elder son, and it needed experiments on all the dogs in the country to cure him then it would be worth it.

Her anxieties, however, have now become part of a widespread movement. Sometime during the aftermath of 1968 'Save the Whale' badges began to sprout. I used to respond glibly enough when I saw the badges on demos. Well sure, but I'd rather save humans first, and go back to my experiments. But the challenge raised by the animal liberationists has sharpened. Books on animal rights have proliferated, legislation to reform the laws on animal experimentation creaks through parliament, and newspaper ads for and against 'vivisection' slog it out as vigorously as soap powders. The debate cannot be ignored — least of all by those of us whose work and even personal safety is threatened by the increasing militancy. But why has the movement grown, seemingly generating at least as much political fervour as reports of human massacres abroad or mass redundancies at home? Is it an aspect of an enlarged moral consciousness — or the displacement activity of a society unable to cure its own human ills?[1]

What attitudes should socialists have towards the animal liberationist cause and its moral and political claims? Is it just one more good theme to be added to a list of things we support, like stopping biological weapons and getting the contras out of Nicaragua? Or does it make fundamental new claims about how we as humans should relate to the non-human world around us? Must socialists be animal liberationists just as much as they must be part of anti-racist and pro-feminist struggles? And what should I as one whose work directly involves animals do about it? The movement calls me a 'speciesist'. I think its a silly term — but if I must, I'll accept the label.

The labour movement and the animals

By-and-large the traditional labour movement has been relatively indifferent to the concerns of the animal liberationists, as it has to those of the ecology movement — or indeed feminism. Apart from that strand of pacifist, vegetarian socialism which has always argued tht humanity cannot be free while animals are in chains, labour movement socialists have tended to be clear that nature and all its works exist for human delectation and domination. Achieving socialism, whatever we hope it will do to relationships between people, is seen to demand a technological abundance, produced by a science which will master, control and exploit the natural world.

As for the scientists — the 'vivisectors' who figure as the villians of the animal liberationists' script — their reactions have been predictably hostile. True, some physicists could be heard breathing audible sighs of relief that it was now *biologists* who were under scrutiny. But the biologists — physiologists, pharmacologists and psychologists who do the bulk of animal experiments — have been bemused, troubled and angered, and sometimes frightened, by the attention and hostility their actions have begun to evoke.

Their arguments are straightforward enough. In Britain, whatever may be the situation in other countries, animal experiments can only be done under Home Office Licence. Conditions under which animals are held tightly regulated and labs are regularly and without warning visited by Home Office inspectors. All animals used must be recorded by the experimenter who has to make an annual declaration to the Home Office. And the new legislation coming into force in 1987 will be tighter still. Isn't all this enough? After all, some would argue that humanitarian controls are rather greater in proportion than those judged appropriate for the protection of human rights in deportation cases or the sheltering of homeless families. Of course, for the fully developed animal liberationist position, the answer is no. What is at issue is the legitimacy of any experiments on animals at all.

But, the orthodox biologists reply, surely it is clear that the work we are doing is for the benefit of humanity? You wouldn't want toxic additives in your food or cosmetics, would you? And improved medical knowledge and new drugs can only be developed on the basis of experiments made on animals. Even *animals* benefit from new drugs and treatments available to vets. Admittedly, there are embarrassments, like those unfortunate experiments down at Porton looking at how animals heal from bullet wounds, and testing new chemical and biological weapons, but after all the country does need defending. Surely better to use animals than humans?

Listening to such apologias, might lay people not be hearing the slightly weary tones of the experts, explaining to the citizenry that it is alright really, all for your own good; that we know best and that you shouldn't worry your heads over it; just leave it in our capable hands, trust us, don't pry too deeply. The trouble is, of course, this is the same team of experts that has brought

Sellafield and acid rain, and dishes out over 50 million tranquillizer prescriptions a year. What vested interests are they protecting this time? If they've nothing to hide why don't they let the public see inside their labs?

There is no doubt that the arrogance of scientific experts has contributed to the growth of the animal rights movement. Perhaps the expansion of the movement over the past decade can be seen as part of a much larger-scale hostility to modernism and the fragmentation of life that our scientised, capitalist-rationalised world produces. Perhaps too, it is a reflection of the yearning of a largely urban society for a dreamed-of pacific world of nature — certainly it seems to be largely a city phenomenon. Few country people, more directly in touch with non-human animals, seem to share its concerns. Some features of the movement seem uniquely Anglo-Saxon, part of the pecularities of the English. Nonetheless ecological concerns are widespread in many industrial societies. Concern for the protection of endangered species, for the elimination of gratuitous cruelty — to humans as well as non-human animals — may be part of the enlargement of social consciousness that the leisure, enforced or otherwise, of a partially post-industrial society offers us.

The politics of animal rights

From the outside, there seems a wide spectrum of positions within the animal rights/animal liberation movement. At the far fringe, there are those who demand a total cessation of any activity which 'exploits' animals, from farming and eating meat to biological research. Spokespeople for the Animal Liberation Front phrase their strategy in military terms: assaults on the property or persons of those they identify as 'vivisectors', break-ins to laboratories to 'liberate' the animals; attacks on butcher's shops; and of course sabotaging hunts, all come within their terrain. Older established bodies, such as the British Union for the Abolition of Vivisection (BUAV), may adopt as absolutist a theoretical position but draw the line at ALF-type 'actions' involving violence. By contrast, organisations like the RSPCA accept the need for some animal experimentation while seeking to tighten the laws, to restrict what is done.

From their opponents come robust rejections of further controls on animal use, for instance from the drug industry in its expensive advertising campaign, and from the Research Defence Society (RDS). Others try to meet the animal liberationists half way. FRAME (the Fund for the Replacement of Animals in Medical Experiments) advocates the development of alternative techniques, like tissue culture methods, and abandonment of testing procedures such as the so-called LD_{50} test, which is a way of measuring the toxicity of a possible new drug or chemical additive by calculating the dose at which half the population of animals exposed to it would die.

The relationship between animal rights and 'conventional' political views is varied — there is certainly no simple mapping of radicals versus conservatives or reactionaries. The anti-viv lobby in the House of Commons spans the parties, and so do supporters of the RDS. Not many radical scientists are likely to be as enthusiastic as the drug industry for keeping controls on experiments to a minimum, but being 'on the side' of the animals doesn't necessarily go along with being on the side of creating a more just or socialist human society. Extreme right-wing groups such as the National Front have begun to adopt animal rights slogans. I've come across one 'animal-libber' who was also involved in a uniformed Nazi cult, and apparently refused to hold a banner proclaiming 'Animal Belsen' in a demonstration outside a German drug company's labs because she felt the slogan conceded the existence of the concentration camps, which she denied. When I raised this case with the BUAV — which subsequently passed a resolution condemning fascism and racism at its AGM — the organiser agreed with me that she was concerned at extreme right penetration of the movement but then went on to compare me, as a 'vivisector', to Mengele (a tricky comparison for someone reared, like me, as an orthodox Jew).

The reasons for the fascist interest in the animal rights movement is only partly to be found in the cover it offers for campaigns against halal and kosher meat and 'zionist' control of multinational drug companies. It also reflects the deep ambivalence within the ecology movement in Britain to human relations with nature. Whatever the case with Green parties elsewhere in Europe, here the ecology movement has tended to face in two directions.

On the one hand, there has been the progressive but anti-centralist tradition whose goal is a humanity living harmoniously with nature, in a society which has transcended and transformed the dominating, extractive and polluting contemporary technologies in favour of an egalitarian world of self-supporting utopian abundance. This strand also recognises that it is in the nature of living processes and creatures continuously to transform the world around them. Humans or no, ecology is about a dynamic, changing universe. Humans are part of that process of change, and to pretend that the world can be biologically 'frozen' as it is now — or was two or three hundred years ago — is to misunderstand the nature of life.

The opposing strand in the ecology movement ignores this dynamic. It leads backwards rather than forwards, to a belief in the possibility of a more static, pre-industrial world where everything and everyone had a place and knew where it was. On a global scale this strand opposes Third World development schemes which would eliminate the last of the noble savages, happy in their poverty, and destroy the wilderness in which the privileged can now safari, their last refuge from universal Benidormdom. Nationally, its ideal is a Wordsworth poem in a Constable landscape, that eternal English countryside with its 'natural' moors and hedgerows (actually not 'natural' at all, but the products of human technology and politics — the forest clearances and Enclosure Acts) and post-feudal but pre-industrial social relations. For this strand, a 'natural' way of living is one given by such apparent biological universals as the 'natural' inequalities between the classes, between the races, between the sexes. It is the vision of the Council for the Preservation of Rural England. Or of fascism.

'Speciesism'

Whatever the tangled political strands within the animal rights movement, it is undeniable that living creatures are made over in huge numbers to serve human purposes, whether killed by antibiotics, hunted by fishing boats at sea, bred on traditional or factory farms, or in research laboratories. The core of the animal rights argument is that these uses are relatively or absolutely unacceptable. The theoretical rationale has been given most recently by philosophers like Mary Midgely in England and Peter Singer in Australia.[2] Essentially, it is that living creatures — especially animals — have rights inherent in them by virtue of existing, and these are to some extent dependent on the degree to which they can be regarded as conscious or capable of suffering pain; hence their use by humans represents oppression and exploitation; a denial of these inalienable rights. Singer has used the term *speciesism* to signify this oppression, by direct analogy with racism and sexism. Coral Lansbury[3] has even made the parallel more explicit by claiming a link between the motives of animal experimentation, pornography and gynaecological abuse of women.

I believe the concept of 'speciesism' is misguided and unhelpful. To begin with it misses the fundamental ecological point that all living organisms exist in a complex web of interactions, some competitive, some co-operative. No species can exist in isolation, without impinging upon others in the ecological web which binds us all. In this web 'rights' inevitably conflict. The 'right' of the wildebeeste or the mouse to survive is in contradiction with that of the lion or cat to eat; the 'rights' of grass and trees conflict with those of the cows and elephants which graze and browse them; the rights of the fly are in contradiction with those of the spider, of the bacteria with the virus which preys on it, and so forth. Eventually even Singer compromises, and looks for some type of cutoff point which is based on the premise that (though it gets wrapped up in somewhat fancier language) the 'closer' in evolutionary terms a creature is to the human, the more rights it gets — monkeys more than cows, all mammals more than fishes, all vertebrates more than all invertebrates, and every animal more than plants or bugs. This is sensible, but scarcely avoids being guilty of some modified form of 'speciesism' (mammalism? vertebratism?).

But my objection to the term goes deeper than its practical illogicality. Crucial to the concepts of class oppression, of racism and sexism, is the coming into conscious political awareness of their own oppression by workers, by blacks, by women. In the classic Marxist sense, oppressed groups in struggle cease to be objects, and become the conscious subjects, of history. It is precisely these groups in struggle which then call into question the nature of the social formations which oppress them, and in doing so *define* class oppression, racism and sexism. The recognition of such oppressions does not depend on the refined consciousness of the oppressors discovering in the depths of their liberal consciousness that they have really been racist and sexist all along. They — we, for I write as a white, middle-class male — are *forced* into the recognition of our roles as oppressors by the struggles of the oppressed. In what sense then can non-human animals become the conscious subjects of their own history? To claim that struggles by humans for non-human animal rights are comparable to those against class, race and sex oppression is to degrade and mystify these human struggles, to turn them into mere manifestations of liberal conscience. This is why it is mischievous as well as absurd.

Humans and other animals

A much more interesting argument is that, whether or not animals have rights in some abstract sense, humans, *just because we are human*, because we are conscious and aware of the pain we may inflict, have duties towards animals. I believe this to be true. Our cat, which is sitting on my shoulder as I write, is a creature for which, by taking her as a kitten and making her into a pet, we have in an important way accepted responsibility; it would be wrong for us not to feed her or leave her too long alone in the house. The goldfish and garden plants are also our responsibility, but they can manage a bit better without our intervention. It was in this sense too that our pacifist neighbour all those years ago was right to worry away at the question of my experimenting on rats and chicks. Quite apart from what it did to the animals, what ill-effect might it have on me as a person — did it not degrade my sensibilities to kill animals?

There is clearly some truth in this. As a student biochemist I had to learn to kill animals — rats, guinea pigs, chicks, and to dissect out tissues, prepare samples of brain and liver. Dissection is a skill, and there is a lot of satisfaction in achieving it quickly and tidily. But I don't believe any biologist ever kills a lab animal with anything other than a sense of regret — it is not like the way the grouse shooter or salmon fisher go about their sport — or even the way a farmer shoots a rat in a barn. Indeed I'd go further — one couldn't be a good biologist if one had no respect for the creatures one studies. You can see this tension in the strange language that biologists use to talk about killing. Research papers frequently refer to 'sacrificing' animals as if their lives were being committed to the great god of science. I loathe this circumlocution, but I understand what drives people to use it.

The work we do changes the sort of people we are — how we relate to the human world as well as the world of objects. If our work means we put violent hands on other living creatures — as farmers, butchers, surgeons or experimental biologists — we cannot expect to have the same relationships to other humans as do novelists, schoolteachers, assembly-line workers or philosophers. But those whose work does not mean that they are in regular contact with the life and death of animals nontheless rely in their daily lives on the knowledge and products our labour produces. So the important question is whether my labour, which involves working on animals, is in fact necessary.

A popular animal liberationist argument is that it isn't necessary because experiments on animals can't provide useful information about humans. The example they offer is the thalidomide tragedy, where the drug passed its animal toxicity test without its disastrous effects on human development becoming apparent. And they claim that alternative methods, like tissue culture or trials on human volunteers would be more appropriate. Again the argument is disingenuous. What thaliodomide showed was that more, and more subtle, tests on animals are needed before one can be reasonably sure that a drug is fit to be tried on humans — that is if one wants the drug at all — and that's a point I come back to in a moment. Species do differ biologically in interesting and not always obvious ways, but the overwhelming similarity between the biology of humans and other animals means that one can get real information about human diseases and their treatment from studying animals, and often there is no other way.

Tissue culture — growing human cells in a test-tube and studying the effects of new chemicals on them directly — can give some help, but there are many types of knowledge that just can't be got that way; there's no way tissue culture is going to tell you about drugs which affect mood and behaviour, for instance. As a classic example, take the case of diabetes, known for centuries as 'the pissing evile' in which large quantities of urine with 'the smell, colour and taste of honey' are produced. The urine contains sugar, which the patient's body needs, but in the disease the sugar can't get from the blood to the tissues, and so is excreted in the urine instead. No effective treatment existed, and diabetics simply died. In 1889, Oscar Minkowski and Josef von Mehring in Strasbourg found that if they removed the pancreas of a dog, the animal became diabetic. So could something produced by the pancreas alleviate diabetes — in dogs or humans? Simply mincing pancreas up and feeding it to diabetics didn't help. But in Toronto in the 1920s, Frederick Banting and Charles Best tied up the ducts which lead from the pancreas in dogs and collected the juice that was secreted. Injected back into diabetic dogs or humans the juice stopped the sugar entering the urine and the patients recovered. Banting and Best were able to purify the active agent — insulin — from the pancreatic secretion. Their laboratory boss brought the America drug company Eli Lilly into collaboration and insulin, purified from pancreases obtained from slaughter houses (now derived largely from pigs), has been used in the treatment of millions of diabetics for the last 60 years.

Now I do not see how the biochemical origins of diabetes, and the role of insulin, could ever have discovered without experiments of this type — even with today's technology. Because insulin only works in more-or-less intact physiological systems, tissue culture approaches, or studies on human diabetic volunteers, could never have given the answers. However, there is a modern twist to the story; genetic engineering techniques — themselves dependent on animal experiments — are beginning to provide a replacement to slaughterhouse pancreases for the insulin on which the world's diabetics depend.

Such examples can be multiplied many times, and unless you value dogs and pigs more than people, the conclusions are obvious — at least for a speciest like me. Of course, there are complexities to the story — the role of big drug companies, the question of how much diabetes would occur if we changed our food policy, dietary practices and so forth. But the main point remains. This is why I don't believe there can be a properly consistent, absolutist animal liberationist theory. It's always a question of compromise and ad hoc moral decisions. For instance, my guess would be that a fair number of animal liberationists would also be in favour of pro-abortion legislation and of 'a woman's right to choose'. But what is the logic of offering *animals* rights and then denying them to potential humans aborted when their level of brain development is certainly that of many vertebrates? The point is that there is a conflict of interests between a woman and her foetus in such cases, and we must be clear that in supporting a woman's right to an abortion we are also supporting her rights *against* those of the foetus.

Socialism and the uses of animals

Where does all this leave socialists? Despite the problems of the absolutist animal liberationist position, it expresses the strong revulsion felt about the excessive, unnecessary and often obnoxious uses of animals for human ends which must carry force. So too must the insistence of the progressivist strand within the movement — and within the ecology movement — that we need to develop a science and technology that is in a harmonious rather than a dominating relationship with the natural and animal worlds around us.

Let's try, then, to get away from the rights argument with all its problems. The language of socialism is, as we are often told, the language of priorities. How might this help us? To answer this we need to know what animals are actually used for. For instance, some 38 per cent of all experiments done on living animals in Britain are classified as 'for the selection of medical, dental or related substances'. Let's agree that we could save some of the 1.6 million animals used each year under this heading by adopting alternative techniques like tissue culture, by modifying the LD_{50} test and so forth. My guess is that such reforms won't change the numbers very much. But the real question to ask is what do we want all these new drugs *for*? According to World Health Organisation figures, there are already some 60 000 branded drugs and other medicines on sale — yet only a small proportion of these (WHO estimates

220, but the figure is probably higher) are considered necessary, well-documented drugs for well-documented disorders.

The main thrust of drug research by the pharmaceutical companies is to find ways of maintaining market share and profitability. To do this they are continually in need of new formulations which circumvent patent laws and 'me-too' agents, so that each major company can offer a range of medicines across a wide area of human disease and discomfort. It is to this end that the overwhelming majority of that 1.6 million animals a year are being used, whatever the disinterested pretensions of the drug companies in their advertising campaigns.

But the argument goes further. Drugs are used in the treatment of human illness and disease. Yet we live in a society which *produces* such disease, which multiplies cancers by the use of tobacco and environmental carcinogens, which encourages the unhealthy eating and living which helps produce heart disease, which creates environments and stresses which lead to anxiety and depression. A socialist strategy for health would be one which started at the right end by endeavouring to eliminate the social causes of ill-health. Drugs are not needed in such a task, and nor are the animal experiments which their production demands. I'm not arguing that we could ever produce a society in which there would be no ill-health and no need for drugs; our biology is also responsible for our health and disease! But clearly, if we could get our priorities right, we could reduce enormously the scale on which such animal experiments were required. But never to zero, and it is disingenuous to pretend otherwise.

So where does this leave the sort of experiments I do? For I am a sort of test case for a good bit of animal experimentation too — part of that area of research sometimes called 'basic'. I am trying to discover some of the biological brain mechanisms involved in the workings of the mind — the processes of memory, and the ways in which, when we or other animals learn, the new information is stored as coded 'traces' in the cellular structure of the brain. There is no way that this type of question can be answered without working with some real animals — training them on various tasks and analysing their brains — though there is a possibility that some experiments can be done with rather simple creatures like sea slugs, which presumably raise fewer problems for most animal-righters.

Now I can't honestly claim that answering these sorts of question will lead immediately to improvements in human health or welfare — although I hope we may learn a good bit about how our own minds and brains work and how they develop during childhood. And it is in the nature of research that no outcomes can ever be fully predicted. So should such experiments be done?

Those who say no might argue in two ways: they might claim that if we really want to know about human learning and memory we are better off reading novels or studying human psychology than analysing the 'hardware' of the brain. There is a long-standing debate between such holistic and reductionist approaches in psychology, as there is in many areas of biology;

my own view as argued in other chapters, is that we need both types of understanding if we are to approach any sort of meaningful understanding of the material reality of the world around us and our own place in it.

The second argument is one about priorities: answering questions about the mechanisms of memory may be all very well, but perhaps there are many more urgent issues to which my biological skills ought to be addressed? I think this is probably true; a socialist Britain might well adopt a science policy which set a number of urgent national goals — from improving the perinatal mortality rate in working-class babies to diminishing mental retardation, or the excessive deaths from coronary heart disease, or finding a strategy to reduce the sky-high psychiatric casualty figures which the way of life of modern industrial society seems to generate.

For such goals, I and many other 'basic' biomedical researchers could well be mobilised. But we don't yet live in a society capable of setting such goals, and one of our political tasks is surely to help build one. This is part of what making a science for the people is about. In such a society, decisions about what types of research to do — whether directly to contribute to solving some immediate problem, or to long-term, more 'basic' goals — could and should be made openly and democratically, in a manner which involves working scientists and 'lay' people. Experiments which might involve hazards, or utilise animals, would be subject to very careful scrutiny. Only if the knowledge they were likely to provide was judged helpful and non-trivial, would they go ahead if they involved causing pain or killing of animals — especially animals with complex brains and nervous systems like mammals.

Even in such a society though, room must be found for the creative and not immediately 'goal-driven' exploration of nature; indeed in a truly socialist society there would be more space for such work just as there would be more unlocking of the creative potential of writers, music makers and painters. In that sense, the gaining of understanding about the world, by the methods of science as well as other cultural forms, is a fundamental part of being human. I'm not sure that scientific explanation of the world is 'in our nature' in the same way that it is 'in' a cat's nature to hunt and kill mice. I do know that it is a part of being human that I would struggle to defend. If this is 'speciesism' then I accept the label. Continued human existence and happiness inevitably means both co-operation and competition with other life forms. We should maximise co-operation, minimise competition, and learn to value non-dominating relations with the natural world. But where there is conflict, as there must always be, I have no doubt that my primary loyalty is to my fellow humans.

Notes and References

1. Particular thanks to Jane Bidgood, Lynda Birke, Maggie Boxer, Ben Rose, Hilary Rose, Dawn Sadler, Gail Vines, Steve Walters ans Stuart Weir for comments on several drafts of this chapter, but I should emphasise that the responsibility for all views and comments in it is mine alone.

2. M. Midgely, *Animals and Why They Matter*, Penguin, 1983; P. Singer, *In Defence of Animals*, Blackwell, 1985. See also S. Clark, *The Moral Status of Animals*, Oxford, 1984, for the 'liberationist' case. The orthodox — not to say a trifle complacent — scientific case for animal research comes from W. Paton, *Man and Mouse*, Oxford, 1984.
3. C. Lansbury, Old *Brown* Dog: *Women workers and vivisection in Education England*: U. Wisconsin Press, 1985.

7 From Causations to Translations: A Dialectical Solution to a Reductionist Enigma

Mind and brain

What is the relationship between the biochemistry and physiology of an organism and its behaviour, whether as an individual interacting with its physical environment or as a member of a social group in interaction with its peers, parents or offspring? The 'brain and behaviour sciences' have become one of the largest organised research areas among the 'basic' sciences in all advanced industrial countries. The Society for Neuroscience is now one of the biggest professional scientific groupings in the USA, for instance. Traditionally, many different disciplines have sought to understand brain and behaviour – ethology, psychology, physiology, pharmacology, and more recently, biochemistry and molecular biology, to say nothing of cybernetics and mathematics. In the past each discipline has tended to be privatised, to talk only to itself, to try to 'complete' its own world-picture.

At the same time the philosophical and epistemological claims of the different fragments of knowledge into which bourgeois culture has split our understanding of the world are greater than this. As I have shown in earlier chapters, the dualism of Descartes gave way during the eighteenth- and nineteenth-centuries to a full-blooded mechanical materialism or reductionism and radical physiologists embraced a crude 'mechanical materialism.[1] This was the mechanistic argument against which Marx and Engels inveighed, seeking to counterpose to it a dialectical understanding of the relationship between different levels of phenomena. The incipient dialectical materialist approach to what philosophers persist in calling the 'mind/brain problem' achieved a brief flowering in the Soviet Union during the 1920s and early 1930s but was hammered into submission by a crassly mechanical reading of Pavlov which simply reduced the psychological to the physiological and dressed it up in Stalin's version of dialectical materialism, or 'Diamat'.[2] The dialectical endeavour has not subsequently been developed, and in the meantime in the West, as I point out in Chapter 2, reductionism has become the dominant mode of interpreting the world.

Bourgeois philosophy, which for a long time held out against the reduction of mind to nothing but a ghost in a machine, the whistle of the steam train, has come increasingly to accept this reductionism at its face value. As identity theory, in the hands of Place, Armstrong or the 'United Front of Sophisticated Australasian Materialists',[3] it is now probably the dominant 'solution' to the 'mind/brain problem' in the West.[4] Against this reductionism on the part of molecular biologists and philosophers, there has been counterposed only a rather bizarre coterie of elderly neurobiologists who, dissatisfied after a lifetime of planting electrodes or training rats to press levers, have turned instead to the eclectic holism of general systems theory[5] or various forms of resurrected dualism[6] designed to re-establish the ghost once more. Marxists, at least those whose writings are accessible to me, have tended to ignore the whole issue, partly perhaps as an aspect of the contempt which most Western Marxism today shows towards the natural sciences as a whole (the other side of the coin of the exaggerated respect shown in the 1930s, maybe? — an honourable exception being the Lukacsian tradition developed by Heller).[7]

Yet if we are to combat biological reductionism without relapsing into its obverse, a full-blown sociological reductionism which denies the ontological unity of the world entirely, and is now an increasingly fashionable escape route both among bourgeois[8] and libertarian[9] writers, we must endeavour to come to terms with this question.

On levels of analysis

I must pause for a moment at the outset and say a little more about the question of 'levels', first discussed in Chapter 2. It is a commonly enough used term — by dialecticians as well as reductionists — yet one whose implications are often a bit taken for granted. In dealing with the sciences of complex systems like living organisms, it is generally assumed that there is a hierarchy of orders of analysis, from the physical to the social. Every level of the hierarchy corresponds approximately to the boundaries of one of the traditional scientific disciplines: physics, chemistry, biochemistry, physiology, psychology, etc. Levels are thus essentially operationally and epistemologically defined; clearly they are not ontologically distinct and their objects of study may be identical. Despite the fact that the rules of experiment and proof, and the language of explanation of each level acquire a certain autonomy, the purposes of studying an object at one level may only become apparent in the context of the other levels at which it may be approached. Nevertheless, the hierarchy of levels is not considered symmetrical; it is given an upwards and downwards direction. 'Upwards' is seen as in the direction of increasing complexity, 'downwards' traditionally in the direction of increasingly 'fundamental' or 'basic'. (One problem with the use of the term 'level' is that the upwards and downwards directionality is particularly conducive to reductionist thinking. In continuing to use the term here I do not in any way mean to give *ontological* priority to the lower levels.)

Now it is clear that, at least at the present time, the levels of analysis in the hierarchy correspond to a social division of labour among researchers. Although there is some communication between levels, which is the subject of much of the discussion which follows, workers within each level strive towards a certain epistemological completeness within their own level, despite the more-or-less contested claims that in due course the totalised project of 'the sciences as a whole' will succeed in integrating them in an as yet unspecified way. The debate is over the nature of this integration, and it ranges from a hard-line reductionism to emergentism. Reductionism, as I have said, claims that in the long run higher-order levels will be collapsed into the lower-order ones; that they are 'nothing but' particularly complex systems to solving the problems of which the equations of the physicist are slowly approaching. Emergentism claims an ontological distinction between levels; either that the methods and laws of the lower-order disciplines are of their nature insufficient or inappropriate to deal with higher-order phenomena or that the higher orders, such as the phenomena of social existence and especially of the human condition, are largely or entirely divorced from the lower — the position, I believe, of much of the writings of structuralism, existential psychology or psychoanalysis.

It is not simply that as one ascends the hierarchy the integral objects of study of the lower levels become the compositional units of which the higher levels are perceived as being built; but also that new organising relationships emerge between them and in addition, an open system at one level becomes a closed component of a higher-level system (for instance physiology requires not merely the chemical and dynamic relationships of biochemistry; the spatial relationships between components, studied by anatomy and morphology, are also necessary). It is in this sense that higher-order levels are more complex. The physiologist may study the properties of a single cell, the biochemist the averaged properties of a million such, ground up with a homogeniser; and yet the complexity of the former, the order given by the relationship of its parts, is greater.

This is the 'orthodox' solution within the dialectical tradition and I wish to add to it that the interpretation at each level requires concepts which are themselves only appropriate to that level; *genes* are not spiteful or altruistic, *assemblages of cells* cannot learn, love or be angry; such terms are inappropriate to the genic or physiological levels of analysis, but appropriate to those of the whole organism.

The subjects/objects of study

Many sociologists — even non-Marxist ones — bristle at the use of the term 'behaviour'. They rightly point to the tarnishing of the term with the aridities of Skinnerian behaviourism, the way that the whole tradition of *non*-introspectional psychology of the twentieth century — with its concern for objective measures of verbal utterances or motor activities — ignores the domain of thoughts and emotions unexpressed in action, and to the obvious fact that many different 'internal states' can lead to apparently identical external actions. At any event, in this chapter I will continue to use the term 'behaviour' for both internal and external states and activities of the organism, though I am aware of the dangers of over simplication, and misinterpretation which the use of the concept of 'level' brings.

The inadequacy of systems approaches

Clearly the study of 'behaviour' in this sense itself takes place at many levels. A population can be studied both with a view to following its interactions with the rest of the world and its evolutionary unfolding in time, and with a view to the relationships, successes and failures of individuals within it (the domain now beginning to be colonised by population genetics and sociobiology).

The systems approach to the study of the behaviour of individual organisms[10] is an attempt to provide a 'complete' description of the behaviour of an organism without recourse to the cellular or biochemical actuality, hence it is non-reductive in the sense in which I use the word here.

According to the systems view, organisms are seen as essentially rather sophisticated computers which must possess certain types of functional component in order to achieve specified outputs. The behavioural outcomes are specified in terms of the products of a number of rather arbitrary internal black boxes interconnected in specific manners. The task of research then becomes largely that of observing relevant behaviour patterns and attempting to segment them into units — behaviourists used to see these as particular patterns of motor activity. Today they are more widely regarded as goal-directed functions which may employ a number of possible routes to a particular goal, be it food, orgasm or whatever. The test for this approach is whether the explanatory black boxes can generate predictions as well as *post hoc* accounts; hence the enthusiasm for computer modelling.

What then, is the problem with this modelling approach by which the organism becomes essentially emptied out into abstract function boxes and flow charts? If it can generate explanations and predictions, is this not sufficient? Despite its popularity, modelling of this sort is in my view always flawed, being able to pass the theory test perhaps, but never the reality test.

An infinite number of models is always possible for any given outcome which is being modelled; furthermore, a test of a model which does not bear out its predictions can always be overcome by a modification of the model without a change in its basic elements. After all, the Ptolemaic system of planets rolling in epicycles on the crystal globes of the heavens actually fitted the data as well, or better, than did the Copernican one; it just happens that the world is in *reality*[11] not arranged as the Ptolemaics modelled it. There was no possible way of discomfirming the Ptolemaic model *merely* on the basis of better observations of the transit of Venus of the perihelion of Mercury or whatever; it had to be done by asking quite different sorts of questions about the way the planetary and stellar worlds are organised. The reality test in biology asks (a) Do the processes occur or are they arbitrary reifications? (b) Do the genes occur? and (c) What is the relationship between the expression of particular genes in particular environments and the behaviours under study? It is here that the model systems of ethology and sociobiology come face to face with the real world, and must accommodate to it or become merely vacuous.

The hazards of reification

Science proceeds partly by attempting to identify similarities and differences. Thus it examines disparate phenomena and asks: first, are there common features which underlie and explain them all? And second, what is different about the underlying features which then become expressed in surface differences? (This is what a whole area of genetics is about, for instance.) The problem is that when one extracts out of the blooming, buzzing confusion of the world some hypothetical common underlying feature, say *x*, then this *x* tends to take on a life of its own. Its material reality, rather than hypothetical nature, becomes taken for granted as the paradigm for constructing future experiments; its use becomes extended as a way of explaining quite other phenomena from those for which it was first hypothesised, and so on. Soon *x*, from being a particular *ad hoc* extractive hypothesis, becomes the object in terms of which others must be explained. Think for instance of the way in which falling bodies are said to obey the law of gravity, as if the *law*, which is a theoretical construct, had primacy over the *objects* whose properties and whose relationships it is defining.

Much behavioural research proceeds in this way; *x* may be perception, learning, hunger, sexual drive or whatever. The implication that there is a single underlying, reified thing, *x*, runs the danger of confusing thought instead of producing the looked-for clarifications, as with nineteenth-century phrenology (bumps of philoprogenitiveness, love of music and so forth) or contemporary 'altruism', 'aggression' or whatever. Disparate phemomena may become quite inappropriately grouped in such classifications.

There is an even greater hazard involved in reification, however, and that is when a dynamic process itself becomes turned into a thing, with something of the weight of a material object about it. Behaviour patterns are dynamic interactions of the organism with its environment. One really shouldn't speak of 'altruism' or 'intelligence' as *nouns* describing objects, but as *verbs* describing activities, they do not reside as static lumps inside the organism, to be drawn as a system-modellers box or mathematicised as a population geneticist's phenotype. It is interesting that Richard Dawkin[12] comes close to a recognition of this problem when discussing altruism. Genes for altruism, he says, are any which confer upon their possessor the property of acting 'altruistically' in particular circumstances; thus 'genes for bad teeth' in carnivores would result in their reducing their share of any common food resource to the benefit of others, and by definition this reduction is altruistic. Yet to define 'bad-tooth genes' as 'altruistic genes' is surely to further empty the reified abstraction of 'altruism' of useful explanatory power.

The reductionist fallacy

To the biochemist, reductionism is not second but first nature. Not merely is there no question but that explanations must be bottom-up — DNA determines primary protein structure, determines the folding of protein chains and their enzymic activity, determines cellular architecture, determines function, determines organismic behaviour; but furthermore, many believe that there are not even multiple pathways, redundancies or flexibilities to higher order outcomes. 'A disordered molecule causes a diseased mind' was the dictum of a respected biochemist at a conference on learning I attended in the USA in 1978. (And what is this but a version of Crick's 'central dogman' writ large?) Particular molecular events, the claim runs, 'cause' particular behavioural events. Just as a recessive mutation results in the production of a haemoglobin molecule with a particular single amino acid substitution and this in turn is able to alter the kinetics of oxygen binding to the haemoglobin molecule, 'causing' the disease known as sickle-cell anaemia; and the phenylketonuric mutation 'causes' an enzyme deficiency in the pathway of phenylalanine metabolism which 'causes' irreversible mental retardation, so changed levels of neurotransmitter metabolism in the olfactory bulb of certain strains of mice 'cause' agressive behaviour. Learning is 'caused' by the synthesis of particular protein molecules in the cortex; changed dopamine levels 'cause' schizophrenia in humans, and the possession of an additional Y chromosome 'causes' unusually violent males. If these events were *not* causal, how does one explain the effects of drugs, or justify publishing academic journals with titles permuting the word 'psychopharmacology'?

In this reductionist approach, it follows that the proper task of the sciences of the organism is to collapse the individual's behaviour into particular molecular configurations; while the study of populations of organisms comes down to the search for DNA strands which code for reciprocal or selfish altruism. Paradigm cases of this approach over the last decade have been the attempts to purify RNA, protein, or peptide molecules that are produced by learning and which 'code' for specific memories; or the molecular biologist's search for an organism with a 'simple' nervous system which can be mapped by serial electron microscope sections and in which the different wiring diagrams associated with different behavioural mutations can be identified.[13] If the structuralist and modelling ethologists claim that everything lies in the programme and the hardware is irrelevant, these molecular approaches certainly mistake the singer for the song.

The paradoxes that this type of reductionism gets itself into are probably more vicious than those of the systems modellers. They have been apparent, of course, since Descartes, whose reduction of the organism to an animal machine powered by hydraulics had to be reconciled, for the human, with a free-willed soul in the pineal gland. As then, so today, mechanistic reductionism forces itself into sheer idealism before it is through. In its modern form one finds it in molecular biologists whose ultimate ethic becomes the search for absolute scientific 'truth',[14] in physiologists such as Eccles for whom indeterminacy creeps in through the synaptic clefts,[15] and in sociobiologists where genetic reductionism is always rescued in the final chapter by human consciousness and 'our' free will galloping over the hill in the nick of time like the US cavalry. This ploy is used in for example the concluding paragraphs and chapters of Dawkins,[16] with his 'memes', or Barash,[17] and Wilson,[18] who allow that genetic determinism may, if 'we' wish, be overturned by conscious human effort — the resurrection of an almost Nietzschean 'will'.

Springing the trap

The easy ways out, if one abandons either dualism or emergentism of the forms defined above, are:

(1) To cut off dialogue altogether; each cobbler will stick to his or her last and history will show who produces the better shoe. This may not be entirely negative. Anything else could be seen as premature and possibly even prejudicial to individual disciplines.

(2) A pragmatic Anglo-Saxon solution which binds all the contradictions together with a dash of common sense and a hearty reef-knot. This pragmatic approach is the one that allows 'causes' to run both ways through the reductive chain: molecular events can 'cause' behaviour, and behavioural events or changes in the environment of the organism can 'cause' molecular events; learning 'causes' the synthesis of particular proteins which modify synaptic structures such that when these cells are reactivated they 'cause' recall. The chain begins and ends in behaviour and causes run symmetrically up and down it, presumably passing one another like commuters on parallel escalators.

(3) Another escape is to classify phenomena as of two types, caused *either* biochemically *or* behaviourally, as in the classic psychiatric distinction between organic and functional psychoses. The two types of phenomena are believed to be able to coexist or even interact, as when diseases are 'real' or 'psychic' or 'psychosomatic'. In any event, such causes are seen as part of an exhaustively additive category which follows the algebra of biometrics, where variance is always x per cent genetic, y per cent environmental with $x + y$ approximately equal to 100 per cent. (If there is an interaction between genes and environment, x times y, then this term is always small compared with the others; not, incidentally, because it has empirically been shown to *be* small, but merely because the algebra doesn't work out if the term gets too big, and nothing in nature should be so complex that simple algebra won't solve it).
(4) The pragmatists' final escape route is to abandon cause in favour of correlation. The difference between correlation and cause is supposed to be that cause says that 'If A then B follows' in a transitive and time-dependent way. A is a necessary and sufficient pre-condition for B. A sharpening of this definition would also demand that A was an *exclusive* precondition for B, that is, that it does *not* result in some other phenomenon (B_1) of the same general class as B. By contrast, correlation merely says that when A occurs there is a tendency, other things being equal, for B to occur also. Correlations do not necessarily imply transitive relations (if A, then B; if B, then A); they are probably necessary but need not be sufficient or exclusive (correlations between the rise in the issuing of television licences and deaths from coronary heart disease in the 1950s in Britain are examples of those that are, one assumes, not even necessary but merely fortuitous). So the literature becomes full of correlations between enzyme levels in the olfactory bulb and mouse-killing behaviour in the rat, dopamine levels in the midbrain and schizophrenia in humans, exposure to training situations and enhanced protein synthesis in the chick, or whatever.

Nothing wrong so far, at least I hope not, as the last could be a precis of the title of more than one of my own papers! The trouble begins when correlations are seen as sorts of 'soft' causes, that when we have observed enough cases in which enhanced excretion of phenylalanine is associated with mental retardation, sickling blood cells with anaemia, alcohol ingestion with drunken behaviour or cigarette smoke with lung cancer, we will be able to turn the intransitive correlation into a transitive cause; phenylketonuria 'causes' mental retardation, haemoglobin substitution 'causes' sickle-cell anaemia, alcohol 'causes' drunkenness, and cigarette smoking 'causes' cancer, although we may like to evoke a *ceteris paribus* clause, and it won't stop some determined people trying to show that both lung cancer and smoking have a co-varying prior genetic cause.

From causes to translations

It will be seen that I am arguing that most conventional attempts to relate biochemical to behavioural events, while they may be very good at generating interesting data or even effective interventions into behavioural processes, do so at the expense of ignoring the conceptual difficulties which underlie the task that they are attempting. I am not sure that I am able to provide a rigorous account of the alternative, but I think I do know where the problem lies and the direction in which its resolution should be sought. The theoretical confusion lies in the multiple uses of the word 'cause' when applied to the task of relating observations at differing levels of complexity and analysis. Everyone knows what is meant by saying that depriving an animal of food causes it to seek food more actively, or that tasting an aversive substance causes it to avoid that taste subsequently. Analogous to these are such biochemical statements as 'the addition of iodoacetate inhibits enzymes of glycolysis causing an accumulation of glucose' or 'the addition of mRNA to an appropriate cell preparation causes the synthesis of protein to commence'. These causes are indeed transitive in the way discussed previously and they can be completely contained within a single level of analysis, be this the biochemical or the behavioural. But such causal statements are not of the same type as those which cross levels of analysis: 'learning a taste aversion *causes* enhanced protein synthesis in the chick fore-brain' or 'inhibiting brain protein synthesis *causes* a failure of memory storage,' or even 'the altered flux of Na^+ and K^+ ions across the nerve cell membrane *causes* the axon potential'. These statements describe a different type of relationship.

For reasons first discussed in Chapter 2, although these latter statements may *seem* like causal ones, a little thought shows they are not. If 'causes' can go 'up' a series of levels in a transitive way, they clearly cannot also run 'down' the levels in a similar transitive way without paradoxical consequences. Similarly, the temporal element is not of the same form as in the first set of statements; indeed in the last example it is lacking entirely; there is not *first* a passage of ions across the cell membrane and *then* the action potential; rather the events are simultaneous. Further, they are indeed *identical*; the passage of ions no more *causes* the action potential than the action potential *causes* the passage of ions; there is just one single phenomenon, studied at different levels of discourse. Here we begin to see the clue to the solution to the paradox. Just as 'gene' — at least in one of its multiple usages — and 'length of DNA' are two names for the same phenomenon, so are 'action potential' and 'passage of ions across the cell membrane'. Only the language differs: one is physiologese, the other biochemese. So we may also hope to advance towards translations of 'memorising an aversive response' into 'modified synthesis of particular proteins in particular ensembles of cells', etc.

It would be possible to produce a coherent description of a nervous system in terms of its biochemistry, of regional (statistically described) concentrations of particular molecules and the dynamics of their interactions. Inputs into this system are the fluxes of molecules from outside, glucose, amino acids, oxygen, hormones; outputs are carbon dioxide, urea, water, other hormones, etc. Regional concentrations of molecules and metabolism also vary within the system as a consequence of spatiotemporal events; an influx of acetylcholine into one compartment of the system results in consequent ion fluxes in a slightly different one; substances diffuse or are actively transported from region to region (for instance, down nerve axons, along dendrites) and the rules which regulate such interchanges can be studied rigorously.

A similar physiological model could be produced: the statistical electrophysiological state of the system and its regions could be mapped and changes associated with inputs from sensory nerves, outputs through motor ones, and intraregional synaptic transmission could be described. As with the biochemical description, causal relationships between inputs, changed global and regional states and outputs could be developed. Note that *neither* the physiological or the biochemical description could provide interpretations of the 'reason' for particular inputs.

Now for the relationship between the biochemical and the physiological description. I argue that between the two there is not a causal, but a *mapping* relationship; that is, that there is a correspondence between descriptions of events and processes in the language of physiology, and that of biochemistry. A complete translation of physiology into biochemistry requires, of course, not merely a correspondence of temporal sequences of events, but also a spatial correspondence: that is why I stressed in the first section of this chapter that physiology translates into biochemistry plus morphology. In one sense the mapping that I am describing is isomorphic, although not necessarily one-for-one (as opposed to one-to-many or many-to-many), and it is no more a causal one in its relationship than is the relationship between, say English and French. The two languages can be translated and are generally isomorphous in so far as they are descriptions of the same unitary universe, but it is not possible to claim a reductive primacy for one language over the other.

The language translation analogy also indicates the limitations imposed on our attempts to move between physiology and biochemistry. While certain types of statements and communications can readily and isomorphically be translated from one into the other, others cannot. Poetry is an example; it has meaning within any one language which an isomorphous translation destroys, because its meaning is context-dependent — that is, it relies on a broader knowledge of the particular history of a language to convey its full meaning to a listener. So too with the biochemistry/physiology relationship. Certain understandings of the *meaning* of an action potential, say, are only obtainable in physiologese, certain understandings of the *meaning* of conformational changes in membrane proteins associated with ion movements only in biochemese. While anger, love or learning have their physiological translations, these translations do not have the same context-dependent meanings as in the language from which they have been translated (if they did, it would become possible, if cumbersome, to replace one language entirely by another and we would merely have cunningly reinstated the old reductionism in a new garb).

However, I do wish to emphasise that I am well aware of a particular further problem with the task of translation. Translations assume that there is an agreement about the boundaries of the phenomena to be described, that the English and French sentences refer to the same time, process or event. The development of distinct sciences, of biochemistry and of behaviour, for instance, means that each has built up an internally agreed understanding of what are the natural objects and processes which its discourse addresses. The problems, particularly of behavioural research, are the uncertainty over the status of the objects of its discourse — are they merely reified artefacts or do they correspond to natural phenomena? To even begin the task I have mapped out demands that this issue is clarified with each and every case. At present, I believe there is only a very limited domain over which one can be reasonably sure that the languages of behaviour and biochemistry march in parallel — one promising area, for reasons which I shan't discuss here, but which informs my own laboratory research, is that of learning and memory (see Chapters 9 and 10).

There is one further step to take in the argument. The domain over which it is possible, in principle, to translate biochemistry into physiology or behaviour is further limited. Not all biochemistries have physiological translations; not all physiologies have behavioural, any more than all physics and chemistries have biochemical, translations. Only sufficiently complex chemistry in a sufficiently concentrated spatiotemporal domain constitutes the material for which a biochemical dictionary is possible, and similarly, only sufficiently complex biochemistries make a behavioural dictionary possible. To speak about the physiology of a stone or even a DNA molecule would be absurd; so too would it be absurd to speak of amoeboid psychology. In this crucial but strictly limited way, the relationships between the languages of the different levels are asymmetric. All behaviours of the individual organism must in principle translate into biochemistry (though not all translations will be useful). But not all biochemistry translates into behaviour. It is this asymmetry which enables one to ascribe an order to levels of analysis, to arrange the hierarchy. The higher the order, the richer the meanings and more compact the descriptions; the lower the order, the wider the domain of relevance within the natural world. This may seem to be labouring the obvious, but redefining the theoretical task in this way avoids a number of the experimental and theoretical minefields sketched out in the earlier sections.

While claiming that mind and brain processes are identical, this *dialectical identity theory* insists on the continued legitimacy of mind language. It resists locating ultimate 'cause' in a molecular domain while insisting that the molecular and cellular knowledge is necessary for a full understanding of the material reality of both mind and brain. The tasks of a 'brain and behaviour science', attempting to make a complete dialectical description of the organism and its history and relationships, then become not those of the search for trans-hierarchical causes, but for translations between biochemical events and behavioural ones. Will such a science of translation be liberatory? There is no absolute answer to this; like reductionist science, — indeed, like all phenomena, it will bear its own contradictions within it. All I would maintain is that at this moment in history it provides us with a *better* key to understanding the world than does reductionism; and understanding is, as we know, one part of changing the world.

Notes and References

1. For example, M. Bunge, *The Mind-Body Problem*, Oxford, 1980.
2. L. Graham, *Science and Philosophy in the Soviet Union*, Knopf, 1971. See also A.R. Luria, *The Making of Mind*, Harvard, 1979 and L.S. Vygotsky, *Mind in Society*, Harvard, 1979.
3. For instance D.M. Armstrong, *A Materialist Theory of The Mind*, Routledge Kegan Paul, 1968.
4. E. Wilson, *The Mental as Physical*, Tavistock, 1979.
5. For instance L. von Bertalanffy, 'Change and Law', in A. Koestler and J.R. Symthies (eds.) *Beyond Reductionism*, Cape, 1969, pp. 56–84.
6. K.R. Popper and J.C. Eccles, *The Self and its Brain*, Springer, 1977. A. Peacocke, *God and the New Biology*, Dent, 1986.
7. A. Heller, *On Instincts* and *A Theory of Feelings*, Allison & Busby, 1979.
8. For example, B. Barnes and S. Shapin, *Natural Order*, Sage, 1979.
9. For example, R. Young, 'Science *is* Social Relations', in *Radical Science Journal* 5, 1977, pp. 61–131.
10. For example, K. Oatley, *Perceptions and Representations: the Theoretical Bases of Brain Research and Psychology*, Methuen, 1978. For a critique which I have only belatedly come across, but which is related to mine, see J. Weizenbaum, *Computer Power and Human Reason*, Freeman, 1976.
11. There's a danger about discussing the 'real' world, it runs into criticism not merely from sociological relativists like Barnes and Shapin, and Young (see above), but from the fragmented and uncertain writings of the divided world of the bourgeois philosophy of science, from Kuhn to Feyerabend. But for a defence see R. Bhaskar, *A Realist Theory of Science*, Verso, 1977; H.A. Rose and S.P.R. Rose, 'Against an over-socialized conception of science', in *Communication and Cognition*, 13, 1980, pp. 173–87; and 'Radical Science and its Enemies', in R. Miliband and J. Saville (eds.), *Socialist Register*, Merlin, 1979, pp. 317–35.
12. E. Dawkins, *The Extended Phenotype*, Freeman, 1982.
13. For example, S. Ward, N. Thompson, J.G. White and S. Brenner, 'Electron microscopical reconstruction of the anterior sensory anatomy of the nematode *Caenorhabditis elegans*', in *Journal of Comparative Neurology*, 160, 1979, pp. 313–18.
14. J. Monod, *Chance and Necessity*, Cape, 1973.
15. J. Eccles, in Popper and Eccles, above as well as many of his earlier books and essays.
16. R. Dawkins, *The Selfish Gene*, Oxford, 1976.
17. D.P. Barash, *Sociobiology: the whisperings within*, Souvenir, 1981.
18. E.O. Wilson, *Sociobiology: the new synthesis*, Harvard, 1975.

8 Disordered Molecules and Diseased Minds

On biological markers

When I was asked, in 1983, to take part in a World Health Organisation Symposium on 'biological markers in mental disorder', I was puzzled. I am neither a psychiatrist or medical doctor. I have never sought to study the mode of action of a drug on an animal model of a psychiatric disturbance, and yet I was asked to give one of the opening talks. My claims to legitimacy could only be those of a 'basic' neurobiologist whose research concern is the interpretation of the relationships between brain structures and biochemistry on the one hand and changes in the behaviour of laboratory animals on the other. But the invitation set me thinking, and trying to make sense of the wealth of literature on such 'biological markers'. This has been a fertile field over the past decades for the generation of research programmes, grant requests, hypotheses as to the 'causes' of mental distress, and interpretations as to how agents which alleviate that distress may work. Yet despite the wealth of published papers, the increasing sophistication of biochemical, metabolic and genetic methodology, the synthetic skills of the pharmaceutical industry, and the vast tonnage of prescribed psychotropic drugs worldwide, it is hard to avoid a sense of unease, of uncertainty about the questions which the whole endeavour is addressing. What follows then, is a reflection on that literature from the outside.

The intention is not to replay for the umpteenth time the long drawn-out battle between medical and social models of mental distress. Nor am I a dualist who believes in a split between mind and brain. I am committed to a materialist belief in the unity of the two terms. What does such a glib statement mean? Surely that any event or activity which can be described in psychological, 'mind language' must also have a description in biological, brain language. Thus, the acts of speaking and listening could be redescribed in terms of patterns of nerve impulses, activities in particular sets of cells; so many synapses activated, so much transmitter released, in the brains of speaker and listeners alike. We will all agree that my brain is in a different state when I am talking than when I am not. That different state may be described as a 'marker' for my speaking rather than being silent. With appropriate application of computerised brain scans, electroencephalograms and biochemical monitoring, I might be able to define that difference more precisely.

But what would such a measure entail, supposing one could make it? There is, I suspect, one fundamental limitation to such markers. However good they are, and however good materialists *we* are, we do not expect it will prove possible to translate the *content* of my speech into changing molecular structures or cellular activities. All mind processes translate into brain processes and vice versa, but there is an important way in which the languages of mind and brain are incommensurable. The information content of my speech does not translate into statements about molecules or cells, even though both speech and molecules are aspects of the same unitary phenomenon. Saying 'I love you', 'that man makes me angry' or 'I feel miserable or suicidal' carries sets of meanings which are given by personal history, culture, social and economic circumstances and are not reducible to the mere motion of molecules.

I am labouring this point because, presumably, our search for biological markers for mental disorder is motivated by the belief that to find such markers will help us either to explain, or cure, or at least intervene therapeutically in that disorder. Why else should the World *Health* Organisation be interested in such an endeavour?

The core of the problem is as I have already summarised; a few years ago I shared a platform with a colleague discussing the problem of learning disability before an audience of parents of 'learning disabled' children. We need biochemical research, my colleague stated, to discover 'the disordered molecules which cause diseased minds'. It is just this question of *cause* which needs to be explored in some detail, because therein, I believe, some of our conceptual problems lie.

As I describe in Chapters 2 and 7, there are several distinct ways in which biologists use the word cause, often confounding statements about linear sequences of events — within-level explanations of a phenomenon, and statements of identity or partial identity — with between-levels descriptions of a phenomenon. Contrary to the view of my biochemist colleague quoted above, disordered molecules do not cause diseased minds in the same way as for example, the firing of a frog's motor nerve causes its muscle to twitch, to use the example given earlier. At best such disordered molecules may be considered to be identical with (alternative descriptions for) the diseased mind.

If we want our question about the causes of mental disorder to be answered by an analogous, linear type of explanation — that is, what happened such that a particular individual showed mental disorder or distress — then offering statements about changed neurotransmitter metabolism is no sort of an explanation at all. At best it is merely a translation of the problem from one language to another.

The medical model

Nonetheless, the power of the medical model, the power of reductionist thinking, is such that I would lay odds that most psychiatrists, and virtually all pharmacologists and biochemists, really do believe that the description of the phenomenon of mental distress in terms of the properties of particular molecules or ensembles of cells would be more fundamental, more scientific and more rigorous than the description of the same phenomenon by a clinician, a sociologist or a novelist. Just why reductionism is so seductive lies, I believe, deep within our own social and scientific history. I've hinted at some of the reasons in Chapter 2.

Brains and platelets

Once one concedes the reductionist model, the battery of techniques available in the search for markers is well known. One may analyse accessible body fluids of patients and 'normal' controls and attempt to detect statistically reliable differences in likely metabolites; this procedure is roughly analogous to attempting to find out which way the inhabitants of a house vote at elections by examining their grocery purchases and garbage disposal. Or one may try to study other more accessible body tissues in the hope that their properties may relate to those of the brain. Often used are the platelet cells of the blood, whose embryological origins resemble those of nerve cells, and which bear on their cell membranes receptor sites for neurotransmitters rather as do nerve cells. Changes in the properties of the platelets are taken to 'mirror' those of the nerve cells, though the logic for this is uncompelling.

Animal models

Another experimental method is to develop animal models. Not that animals normally behave in ways which in general can readily be described as directly anxious, depressed or schizophrenic. It isn't as clear cut as being able to study diabetes, infection or wound healing. But there are treatments which may make animals appear to behave in ways which are characteristic of depressed or anxious or even schizophrenic people, and one can then observe more directly biochemical or electrical changes in their brains. Further, we can examine the effect of drugs known to induce or modify mentally distressed states in humans on animals, and argue from these effects to the biochemical 'lesions' in humans. These procedures are clearly fraught with their own doubly reductionist logical problems, of trying to explain the mind events of one highly complex, conscious and social species in terms of the brain biochemistry of another.

The interpretation of drugs

Of all the methods used in identifying putative markers, by far the most common is to argue *backwards* from the effect of drugs. This procedure, described by Bignami as *ex juvantibus logic*[1] is by far the most commonly adopted, as a quick check on the literature of the last decade on the biology of schizophrenia and the affective disorders reveals. The argument goes roughly like this. Drug X is shown — by double blind, statistically controlled, properly authenticated clinical trials — reliably to improve the mood of depressed patients, or decrease the existential communication disorder of schizophrenic patients. If we can then show that drug X interacts with system Y — perhaps neurotransmitter uptake or binding in blood platelets or slices of rat brain, then it is reasonable to assume that system Y is in some way faulty in the depressed or schizophrenic patient, and therefore this fault is the 'cause' of the disorder.

Let me draw examples almost at random from the literature: first a review by Langer and colleagues[2] entitled 'High affinity binding of ^{3}H-imipramine in brain and platelets and its relevance to the biochemistry of affective disorders'. These authors begin their paper by stating two hypotheses about the affective disorders, involving the abnormal uptake, binding and metabolism of two different classes of neurotransmitters, catecholamines and indoleamines. One class of anti-depressant drugs, the tricyclics, inhibits the uptake of monoamines into nerve cells, and this, they argue, is consistent with the hypothesis that monoamines are involved in depression. But not all anti-depressants have this biochemical effect, and there is no obvious relationship between the effectiveness of the anti-depressant drug and the degree of inhibition of monoamine uptake. So, they argue, one should investigate other sites of biochemical action of the drugs. They therefore explore the effects of the drugs on the binding of radioactively labelled imipramine, a tryclic anti-depressant which interferes with monoamine uptake. They prepare platelets from healthy volunteers and show that imipramine binding depends on age, but is unaffected by the sex of the volunteers. But when they study the binding in 'untreated severely depressed patients' they find that it is lowered, 'suggesting that this binding site might be involved in the biochemical changes related to affective disorders'.

A massive jump to a conclusion. Yet within the next paragraph we learn that the patients they studied included both those suffering from reactive depression (that is, depression ostensibly caused by external events) and uni- or bi-polar endogenous depression (that is, depression ostensibly caused by a person's 'internal state' independent of external events). But there was no correlation between the degree of a person's depression and the degree of lowering of the binding of the drug.

Our ingenious authors are not easily tied down by such a doubt however. 'This might be explained by a cancelling out of two concurrent changes in ^{3}H-imipramine binding'. They claim there is both an effect of the drug, and simultaneously an effect due to the improvement in the patient's condition caused by the drug. And if this won't do then they argue that the decrease in binding sites may be 'a genetic [sic] marker of a suceptibility to depression' — apparently to be studied by looking at the binding of imipramine to platelets obtained from close relations of depressed patients.

What a thicket of *non sequiturs* this experimental logic offers us! If a drug diminishes a symptom, it cannot be inferred that the biochemical system on which the drug acts is the 'cause' of that symptom, or even its brain correlate. If aspirin reduces toothache, it does not follow that investigating the biochemical mode of action of aspirin will cast light on some 'prostaglandin hypothesis' for toothache. Nor does the fact that aspirin will also diminish the pain resulting from a broken leg allow us to deduce that the 'causes' of toothache and of the broken leg are identical.

In the Langer case, we see how failure to find a biochemical difference — in binding sites for ^{3}H-imipramine between different categories of depressed patients — is explained in terms of a common genetic propensity for the different types of depression. What would have happened had a difference been found? A second paper, by Ananth[3] 'Clinical prediction of antidepressant response', tells us. Ananth uses differences in individual responses to drugs to *define* the clinical diagnosis of the disorder. Thus he claims that depressed patients who respond to tricyclic drugs show 'premorbid personality traits of chronic anxiety and obsessiveness', while those who respond to lithium 'manifest characteristics of cyclothymic personality including mood swings, elation and excitation'. That is, patients who show one type of response to a drug are one type of depressive, those who show another are a different clinical class.

Ex juvantibus logic is thus a no-lose logic. It wants to have its cake and eat it. If some groups of patients respond to a drug and others do not, this difference becomes the motor for clinical diagnosis. for instance, depressed patients who respond to tricyclics are regarded as endogenous, those who do not are reactive. If groups clinically described as different *both* respond to a treatment or show a common 'marker', this is taken to mean there is a common genetic predisposition to the disorder. This distinction is sometimes described as between 'trait markers', i.e. differences found between 'normals' and mentally distressed people *whether or not* they are at present actively exhibiting that distress; and 'state markers' which only become apparent when the distress itself is there. Between state and trait, you can scarcely fail as an experimenter, and in either case the molecular explanation rules supreme.

I should emphasise that I have not chosen poor papers of their type to criticise. The authors are respected workers in their field. That is why I worry about the logical and scientific standing of the field itself. Only real conceptual problems could explain the extraordinary range of biochemical markers and hypotheses which have been offered as providing 'the' unitary cause for schizophrenia or for affective disorders over the past decades.

For instance, for schizophrenia, they include anomalous substances in schizophrenic blood[4] (1960s); the dopamine hypothesis (1972 on); the noradrenaline hypothesis (1971 on): the serotonin hypothesis (1955 on); the prostaglandin hypothesis (1977 on); the endorphin hypothesis (1976 on); the amino acid hypothesis (1972 on); the acetylcholine hypothesis (1973 on); the histamine hypothesis (1937 on); the transmethylation hypothesis (1952 on); and several viral hypotheses (1973 on).

At best, such biochemical marker studies offer correlates of behavioural manifestations. Logically, the biological change being studied may precede in time the behavioural state, in which case it may be the correspondent at the biological level of some behaviour/mind state antecedent to the 'disorder'; or it may coincide in time with the behaviour, or it may follow it. Early studies on biochemical markers were led astray by identifying as 'causes' of the disorder substances in the urine which were the degradation products of drugs used to treat the disorder; and we know better than to accept such iatrogenic 'causes' today. But supposing depression 'causes' decreased imipramine binding in the blood platelets just as influenza virus causes increased nasal mucus secretion?

Or, if that appears too trivial a criticism, take a more substantive one. Suppose that a depressed patient has a decreased level of synthesis of a neurotransmitter in a particular brain region. Synthesis of that neurotransmitter will be an energy-demanding process. If less is synthesised, there will be less glucose utilisation in the brain region concerned and this could be measured as a 'marker'. The lowered glucose metabolism is a necessary correlate of the changed neurotransmitter synthesis, but it would be hard to argue that even in the reductionist sense it was the *cause* of the depression.

All investigations of the relationship of brain biological events to mind/behaviour phenomena, if they are to cast meaningful light on the translation between the two types of description of events, need to show rigorously, for each biological process or marker under consideration, that the change in the marker is necessarily, sufficiently and exclusively related to the behaviour being studied. In the next chapter I have tried to identify the criteria that would need to be fulfilled to show this correspondence between biochemistry and behaviour in the case of memory formation, using animal models. The minimal criteria offered there can scarcely be fulfilled even for very simple models of animal learning, although we have gone some way towards it (see Chapter 10). Frankly I find it hard to see how such rigorous criteria can be met in the analysis of mental disorder, granted the problem of animal models and the limitations of molecular roulette and *ex juvantibus* logic.

Even if they could be met, if we could have a complete description of the differences between the brain of a mentally distressed person and that of a 'normal' person, we would still have a statement of correspondence only, it would not be causal in the sense which reductionist thinking implies. The biological description would only become causal in the event that we could show a relationship, not between the biological markers of the mental distress, but between the biological markers and the mind events leading to that distress. One could then map a correspondence of the type:

predisposing mind event	→	mental distress
⇃↾		↿⇂
biological state (a)	→	biological state (b)

This is the logic, presumably, of the search for genetic markers in mental disorder. It must be granted that, even if today's *brain state* does not predate today's mind distress, today's *genotype* must so do. Yet despite the conventional wisdom to the contrary, the evidence for a genetic predisposition to the most frequent forms of diagnosed mental disorder — schizophrenia and the affective disorders — is at best weak. The re-evaulation of the Danish twin and adoption studies by Kamin,[5] Schiff[6] and Lidz[7] cast doubt on what has widely been regarded as some of the strongest data in the field.

The statement that individuals with a particular genotype are more at risk at being diagnosed as schizophrenic or depressed, if verifiable, should not be surprising. Unless all or most of the individuals with that gene show the disturbance, the significance of a 'genetic predisposition' only becomes meaningful in an ontogenetic, developmental context, and it is here that the real difficulty of unpicking the meaning of statements about causation becomes profound. The problem lies in the relationships of genotype and phenotype.

First, the norm of reaction of gene to environment during development means that the phenotype generated by any particular gene differs at different stages of development and in differing body tissues, to say nothing of different environments. Equally, the *same* phenotype can be generated by *different* genotypes. The obvious example is phenylketonuria, discussed in Chapter 4, where the same genotype can generate an irreversibly mentally retarded or 'normal' child depending on the presence or absence of phenylalanine in the diet. However, to return to the immediate point — the fact that the site of the genetic lesion in phenylalanine metabolism is known does not give us any way of understanding which of the multitude of diffuse effects of this deficit on the physiology of the child is a 'marker for' or 'causative of' the mental retardation.

Second, the metaphor of 'environmental effects' on 'genotypic expression' is inadequate, because it defines the organism — and the genes within it — essentially as a passive responder to environmental challenge. Yet, as I have emphasised, organisms do not passively receive their environments, but actively seek them out and transform them. In a paradoxical way this truth has been long understood in the literature of mental disorder. When in the 1930s, Faris and Dunham accumulated evidence to show that incidence of schizophrenia was highest in the derelict inner city of Chicago and lowest in the affluent suburbs,[8] the obvious environmentalist interpretation was rejected on the grounds that it was precisely the schizophrenic personalities (read 'schizophrenic genotypes') who 'chose' to live in the inner city.

There is yet a further problem in understanding the significance of assumed genetic markers for mental disorders. If a disorder is extremely rare in the population, a discussion of genetic predispositions may prove helpful to understanding its aetiology. If it is very common, the genetics seems to be less and less relevant. In the major Camberwell study by Brown and Harris in 1978[9] they found that one quarter of working-class women with children living in the area were suffering from a definite neurosis (mainly depression), whose incidence could be related to significant life events, whereas the incidence among comparable middle-class women was only some 25 per cent of the rate amongst the working-class. Are we to assume that all of these working-class women share in addition a particular genetic predisposition? Clearly the life and environmental circumstances of this group are much better predictors of the likelihood of their suffering depression than any amount of gene-library-making or biochemical analysis. In common parlance, environmental 'causes' are those which ought to be explored if explanations of maximal utility are to be sought. And while the prevalence of the depression diagnosis in women compared with men is anyhow enough to make a simple genetic argument unlikely we should also note that Roy showed that in depressed men, parental loss before seventeen years of age, poor marriage and unemployment were major vulnerability factors.[10]

The question then for those who maintain that there is a utility to the search for biological markers in depression is this: are we being asked to believe that in all of these causes of depression there are going to be found common biochemical abnormalities? That is, is there a unitary phenomenon, depression (or a simple two-or-three subset classification thereof), to which we will ultimately find that there is a matching unitary to two-or-three subset range of biochemical abnormalities? Do all depressed women in Camberwell have, if not common genes, then common levels of catecholamine, indoleamine or imipramine binding to their platelets? And if so, is this not much more likely to be related to depression as nasal mucus is to influenza rather than as PKU is to mental retardation?

There is one further question which the emphasis by psychopharmacology on *ex juvantibus* logic raises insistently in my mind. We can accept that antidepressant drugs do act by changing the absolute levels, ratios or specific binding sites of one or more neurotransmitters in particular brain regions. Yet rigorously conducted clinical trials of such agents consistently show a placebo effect of around 30 per cent, sometimes more, sometimes less, depending on the exact experimental design, the severity of the patient's depression, the duration of the trial and so forth. I may have missed it in the literature, but I am not aware of any biological psychiatrist endeavouring to answer the question: how does the placebo exert its effect?

Social psychiatry can of course refer to such explanations as patients' and doctors' expectations of success, the subtle dialectical relationship of patient to healer. But what do we materialists believe, who hold that to every mind state there corresponds a brain state? Does the placebo also alter neurotransmitter metabolism, and if so, how? If it does not, there can be no specific one-for-one relationship between the depression diagnosis and altered biochemistry. If it does, alas for the specificity of drug-binding studies, and the games of molecular roulette played by the drug companies!

I am critical, but I am not trying to be negative. I said at the outset that in pointing to the logical problems involved in biological marker studies I was not trying to wage again a battle between social and medical models of mental disorder. A unitary materialist understanding of mind/brain relationships, of the sort for which I would argue, must reject such a dichotomy. Just as biological statements about the brain must translate into mind statements, so social, behavioural and mental statements must translate into biological ones.

What this means is that we cannot accept the easy, pluralistic options which divide the word into two sorts of phenomena: *either* X is depressed because of an endogenous neurotransmitter disorder *or* because she has just lost her job or her husband. Psychiatry adopts this electicism too easily; think of the distinction between organic and functional disorders that the standard text books offer. Or look at the way in which right-wing libertarian critics of institutional psychiatry like Thomas Szasz discriminate between states of mind which he believes are 'organically' caused by observable brain lesions, and are therefore the province of medicine; and those for which he claims no brain cause can be found (schizophrenia and the affective disorders) and which are therefore part of an individual's personal responsibility and not that of institutionalised medicine. The logic of Szasz's approach is that, as more and more biological 'markers' for such disorders are found — as they will be — the province of institutionalised psychiatry will expand and the area of personal responsibility decline. My position is very far from this. We have to understand that matters of mind are simultaneously amenable to social and personal explanation *and* biological description, just as the firing of the axon in the frog's motor nerve is simultaneously amenable to holistic, reductionist, physiological, ontogenetic and evolutionary explanations.

Where does this leave the search for biological markers? First, in need of more rigorous criteria for tests which marker-candidates must fulfil. Second, an acceptance that there are likely to be many-for-many correspondences between brain biological events and mental states, rather than the one-to-one or one-to-many which is all that most present psychopharmacological hypotheses allow. This means that we should expect a multitude of biochemical systems to show subtle changes in individuals diagnosed as depressed or schizophrenic. The very magnitude of the changes claimed to have been found in neurotransmitters as a result of the use of the drugs which alleviate mental distress, suggests that they are working more like aspirin in toothache than on the actual site of the trouble, if there is indeed such a site. It is up to the researchers in this area to prove me wrong.

Notes and References

1. G. Bignami, 'Disease models and reductionist thinking in the biomedical sciences', in S. P.R. Rose (ed.), *Against Biological Determinism*, Allison & Busby, 1982, pp. 94–110
2. S.Z. Langer, E., Zarifian, M., Briley, R., Raisman, and D., Sechter, 'High affinity binding of ^{3}H-imipramine in brain and platelets and its relevance to the biochemistry of affective disorders', *Life Science* 29, 1981, pp. 211–20.
3. J. Ananth, 'Clinical prediction of antidepressant response', *International Journal of Pharmacopsychiatry* 13, 1978, pp. 69–93.
4. Note the reductionism of this type of terminology — common shorthand parlance though it be. It is not the blood which is schizophrenic but the individual from whom the blood is derived. By assigning the property to the tissue rather than the person, we legitimate the dominance of the molecule over the phenomenon. See, for example, S.P.R. Rose, *Trends in Neuroscience, 10*, 1987, p. 152.
5. S.P.R. Rose, R. Lewontin, and L. Kamin, *Not in Our Genes*, Penguin, 1984.
6. B. Cassou, M. Schiff, and J. Stewart, 'Genetique et schizophrenie: reevaluation d'un consensus', *Psychiatrie des Enfants* 23, 1980, pp. 87–98.
7. T. Lidz and S. Blatt, 'Critique of the Danish-American studies of the biological and adoptive relatives of adoptees who become schizophrenic', *American Journal of Psychiatry*, 140, 1983, pp. 426–35. See also, T. Lidz, S. Blatt and B. Cook, 'A critique of the Danish-American studies of the adopted-away offspring of schizophrenic parents', *American Journal of Psychiatry*, 138, 1981, pp. 1063–66.
8. H.W. Dunham, *Community and Schizophrenia: An Epidemiological Analysis*, Wayne State University Press, 1963.
9. G.W. Brown and T. Harris, *Social Origins of Depression: A Study of Psychiatric Disorder in Women*, Tavistock, 1978.
10. A. Roy, 'Vulnerability factors and depression in men', *British Journal of Psychiatry*, 138, 1981, pp. 75–6.

9 Towards a Biology of Learning

Learning and memory

The capacities to learn and remember are central to our definitions of ourselves as humans. The investigation of the phenomenology of learning and memory has occupied psychologists intensely and neurobiologists intermittently for a century. The cell biology of the phenomena has been the subject of serious research for a quarter of that time. Apart from the philosophical interest, the attraction of the field to cell biologists must surely lie in the ease with which model systems can be set up in experimental animals in which behaviour can lawfully and adaptively be modified. Such systems are amenable to intervention and analysis, and there is optimism that here, more readily than in any other aspect of behaviour, it may be possible to find a Rosetta stone which will provide the key to translation between phenomena analysed at the behavioural level on the one hand, and at the biological level on the other (Chapter 7).

Convinced anti-reductionists continue to argue that this is impossible; that Proust's madeleine cake will always tell us more about the phenomenology of memory than any amount of training chicks to discriminate colours or rats not to jump off shelves.[1] And it is true that the goal of such a translation still remains elusive, despite the earnest attention of experimenters and the proliferating research reports. The last quarter-century has witnessed three phases in learning and memory research. The first phase (the 1960s), in the aftermath of the initial successes of molecular biology, approached the problem as a simple extrapolation of the genetic code: 'memory molecules', whether RNA or protein (or even DNA) would soon be isolated, sequenced, and the mystery resolved. Reviews of the field in that decade radiate uncritical enthusiasm for the most improbable of experimental results, and researchers rushed into print with data which lacked elementary controls as psychologists embraced the new biochemistry and molecular biologists looked for new fields to conquer. The 1970s were dourer; experiments became more sophisticated, some inflated reputations were pricked and many abandoned the field in favour of something a littler simpler as the tide of funding ebbed away. By the 1980s, the few researchers who remain have collected a corpus of agreed data and even more importantly have developed a methodology. Some of the ill-temper occasioned by early, hasty claims, has begun to diminish and, who knows, we may even become fashionable again. I have a certain optimism, that after a troubled and prolonged childhood our field, if not yet fully mature, may at last be through its adolescence.

Three problems of the field

In this chapter I want to review broad strategies in the study of the cell biology of learning and memory, in order to exemplify some of the analytical issues involved. In the next chapter I will show how a rational research strategy may be followed by reviewing my own experiments of the past few years. I believe there are four major difficulties in this field. I want to discuss three of them here, reserving the fourth for the next chapter.

The first problem is that the phenomenology of learning and memory studied at the psychological level is complex. Do the terms refer to a continuous, unitary process, or are they shorthand for multiple phenomena? Memories are a person's most durable characteristic. During our lives, every molecule in our body is replaced many times over; we may change our appearance through age or disease, loss access to sight or hearing and yet, while we remember, we still exist as individuals. One of the tragedies of the diseases of senility is the loss of personal memory, and hence of who we are. The paradox of memory lies in this peculiar combination of stability and fragility; a childhood memory can perist for eighty years or more; and telephone number or a person's name can be irretrievably lost within moments of being told it. It is clear that memories are in some way 'in' the mind, and therefore, for a biologist, also 'in' the brain. But how?

The term 'memory' must include at least two separate processes; on the one hand, that of *learning* something new about the world around us; and on the other, of at some later date *recalling* or *remembering* that thing. It is inferred that what lies between the learning and the remembering must be some permanent record, a *memory trace* or *engram* within the brain. What form could such a trace take?

As far back as the beginning of this century, when the great Spanish neuroanatomist Santiago Ramon y Cajal showed that the brain was packed with nerve cells (neurons) connected one to the other by way of close junctions (synapses) he speculated that perhaps memories might be stored in the form of changed patterns of connection between nerve cells, making a series of unique pathways. When from the 1920s on it was realised that the brain was in ceaseless electrical activity, the idea that memories could take the form of recurrent electrical signals between cells — so-called reverberating circuits — became popular. But memories can withstand all sorts of shocks to the head or brain, from concussion to electroconvulsive therapy, which totally disrupt this electrical activity, so they cannot be stored in *purely* electrical form. On the other hand, such damage does tend to result in loss of very recent memory, for events minutes to hours before the shock (which is why people who have been in crashes often can't remember the events that led up to the accident.

Hence the idea that memory exists in at least two forms: short-term memory for very recent events, which is relatively labile and easily disruptable; and long-term memory, which is much more stable and stored in the form of changed molecules and/or changed cellular structures and pathways. Not everything that gets into short-term memory becomes fixed in the long-term store; there is a type of filtering mechanism which selects out things which might be important and discards the rest; if that weren't the case, if we remembered everything that impinged upon us, we would soon become hopelessly overloaded. (There are some clinical accounts of individuals who apparently *cannot* filter their short-term memories and store almost everything; they seem to have a very hard time organising their lives).

But in addition to these, well-known, distinctions between short- and long-term memory we have in recent years been offered the antithetic pairs of working and storage; procedural and declarative; semantic and episodic; active and passive; shallow-processed and deep-processed, to name but a few. Further, many have questioned whether learning in the visual, auditory and olfactory modalities, to say nothing of the proprioceptive ones, really involves similar cell biological processes distinguished only the connectivity of the cells concerned — quite apart from the relationship of any of these to human linguistic memory. We don't even know whether memory is saturable, bearing in mind the seemingly limitless capacity of human recognition (as opposed to recall) memory. For the neurobiologist, the question of just what behavioural processes are being studied may not affect the design of a particular experiment, but it matters profoundly to the interpretation of the results obtained.

That is the first difficulty of the field; the second is that the experimental problems at the cell biological level are non-trivial. If cellular reorganisation and molecular reconstruction occurs as the result of a particular 'learning experience' so as to make new pathways, the changes are likely to be small and restricted. To achieve large changes, one would *a priori* anticipate it to be necessary to expose an experimental animal to relatively profound experience; yet the more significant the experience, the more likely it is to involve components (stress, arousal, fear, etc.) whose cell biological correspondents must be controlled for in the experimental design. Under the best of circumstances, repeated biochemical measurements on a population of genetically homogenous experimental animals in matched environments will rarely reduce inter-animal variance below say $\pm$ 5–10 per cent. Hence differences in a measure of less between learning and control animals than 10–20 per cent are likely to be difficult to detect with the degree of statistical reliability which biochemistry and pharmacology routinely expect in order to draw meaningful conclusions.

The third difficulty is that the nature of animal studies of learning and memory demands that the subject learn not adventitiously but in response to externally applied contingencies of reinforcement — of reward or punishment — and later evidences memory for this experience by performing some task or showing some otherwise altered behaviour. Hence the experiences associated with both learning and recall are complex, involving a set of interacting physiological processes both within and external to the central nervous system. All these processes may be *necessary* for learning to occur and memory to be expressed: however many if not most of them may be general-purpose mechanisms associated with arousal, stress, motivation, perception or whatever rather than *specific* to the events of memory formation themselves (if these may be considered, for the moment, as if they were capable of being reductionistically isolated from the melange of behavioural acts occurring in any organism at any time). The weight of this problem is contingent upon the questions being asked at the cell biological level. For research aimed at identifying exogenous agents (for instance, drugs) or internal processes (such as hormonal levels) which improve or impair memory, this may not matter. There have to be some processes in the brain which provide a signal to the animal that the information being received is important and should be attended to and remembered — what has been called a 'now print' signal. Such a signal can be general and independent of the nature of *what* is being remembered. If the research aim is to identify this signalling process, well and good. But if it is aimed at identifying what has been called the 'engram' — the memory trace itself — then more is required.

Even so, the effects of applying an exogenous agent on behaviour may be erroneously interpreted. For example inhibitors of protein synthesis block long-term memory. But one cannot automatically deduce from this that protein synthesis is necessary for memory formation, because when protein synthesis is blocked, amino acids, precursor molecules to protein, accumulate in the brain and may interfere with nervous transmission. So perhaps the effects of inhibiting protein synthesis on memory point to the involvement, not of protein synthesis, but of amino acids?

A rational biology of memory

Faced with these difficulties, how might one place the study of the cell biology of learning and memory formation onto a rational base? I want to begin by asking two rather general, and perhaps more subversive, questions about the endeavour on which those of us pursuing the 'biochemistry of memory' are embarked. They are:

(1) What do we think we are looking for? and
(2) How will we know when we have found it?

The purpose of this chapter is to propose a set of criteria which putative candidates for the titles of 'the' engram must fulfil, and against which the substantial research effort of the past quarter century must be judged. I believe this task to be essential if we are to move beyond a situation in which all biochemical events which occur in the brain of an animal around the time of its subjection to a training protocol are of equal interest, towards one in which we can make some sort of description of the memory formation process which seems coherent in cell biological terms.

Neurobiologists take it as axiomatic that in the living, behaving organism the cell biology of the brain is in constant flux and that, in ways which have yet to be discovered, there is a relationship between these changing brain states and changing behavioural states. But this sentence requires elaboration. Any particular piece of externally observed behaviour may, as we know, be the outcome of many different internal processes. We cannot infer a unique relationship between a particular piece of behaviour and a particular brain state without including in the definition of the behaviour a reference to the particular internal processes involved. Reynolds,[2] speaking of humans, refers to the traid of behaviour/experience/action, an approach which will need to be reintegrated into our study of non-human animals too.

By a cell biological brain state I mean a summary expression of the integral of the biochemical properties of the brain as they are distributed over its cellular space, together with their rate and direction of change. The cell biology of the brain, likc thc arrow fired from the bow in the old Greek paradox, is always simultaneously *somewhere* and *going somewhere else*. This statement of dynamics is not merely a reference to the dynamic biochemical state of brain constituents. At the minimum we must assume that neurobiological changes have both a vegetative and an active significance. It is not merely that the cell biological structure of the brain must develop, be preserved during adult life, and repair itself in the event of damage. Because of the complexity of cellular interactions in the brain these processes, which are relatively without behavioural meaning in, say, the liver, must nonetheless have such meaning in the brain, and this meaning must also impose its own direction on neurobiological change.

That such dynamic changes occur and are related to behaviour may be demonstrated, for example, by observations ranging from the measurement of electroencephalographic changes through computerised tomography to the demonstration of enhanced metabolic rates in specific brain regions in response to particular visual and auditory stimuli. It must be emphasised, however, that such observations are scarcely surprising; it would have been more bothersome if there had been no such metabolic events consequent upon stimulation.

On the other hand, we are very far from being clear as to the relationship between the biochemical state of an ensemble of neurons and the behaviour or experience of the organism containing that ensemble. Cautious authors refer to the changing biochemical state of the ensemble as a 'correlate' of behaviour or experience. Less cautiously, we are sometimes told that particular biochemistries 'cause' particular behaviours or experiences. For reasons discussed in Chapter 7, I would argue that the concept of cellular 'correlate' of behaviour is in danger of being merely vacuous, and I reject the reductionist mode of explanation which seeks molecular 'causes' for behavioural and experiental phenomena. While we can speak of the addition of cyanide to a respiring slice brain tissue as 'causing' the cessation of oxygen uptake, we cannot speak in the same way of a particular molecular or cellular series of events as 'causing' my memory of the opening bars of Beethoven's Fifth Symphony, my son's face, or the formula of benzene. As I have emphasised in earlier chapters, in the case of relations between biochemical and behavioural descriptions, there are not two sequential phenomena, *first* a biochemical and *then* a behavioural one, but a single phenomenon about which different languages, biochemical on the one hand, behavioural on the other, can be used. A given set of molecular or cellular events is (identical to) my recollection of the opening bars of a symphony, but at a different level of analysis. A non-reductive identity theory must be sought to replace the language of cause and correlation which is frequently, loosely and erroneously used in describing the relationship of biochemical to behavioural descriptions of events.

We can thus formulate the following axiom:

Axiom Every behavioural/experiential state of the organism corresponds to at least one specific cell biological state of the organism, most but not all of the relevant cell biology being within the nervous system.

Note that this formulation leaves open the possibility that statements which endeavour to translate from the biochemical to the behavioural may be one-to-one, or one-to-many or even many-to-many. The known redundancy of the brain as a system would suggest that there are multiple sets of perhaps partially, overlapping cellular events which in system terms may mean 'my memory of Beethoven's Fifth symphony', etc. There is also a danger that the formulation of the Axiom may appear too static. Recalling a memory is evidently a process, not a thing, and we must bear in mind that the meaning of the term 'cell biological state' is likely to be dependent on the history of the cells prior to that state. Hysteresis phenomena in physics and chemistry are examples of the importance of the need to understand a system's past history to explain its present properties even in relatively simple systems.

The task of the neurobiologist is to describe the cell biological states which correspond to specific behavioural/experiential states; that is, of learning the translation rules that relate cell biology to behaviour. I call the matching of cell biological and behavioural statements about a given phenomenon, defining *the correspondents* of that phenomenon to indicate this statement of identity rather than causality.

There are problems here which relate to the definition of a behavioural/-experiential 'state', especially when, as is a characteristic part of the methodology of the behavioural sciences, attempts are made to express it in terms of the sum of individual behavioural components, processes, routines, etc. Do these components, processes and so forth have material existence, or are they the reified consequences of the attempt by the experimenter to impose an order upon the world? If we suppose that 'aggression' is a behavioural state or process, we may search for biochemical correspondents for it, and indeed there are quite a few such studies in the literature, But what are our grounds for believing there is such a unitary 'thing' as 'aggression'? Are we not running the same danger as nineteenth-century physics, which postulated an underlying 'ether' which transmitted electromagnetic radiation in space? If we believed in 'ether', we might try to find out what type of chemical or physical properties it had. Yet when relativity theory dispensed with the ether such a search for its chemistry was rendered void.

The problem is that 'meanings' are given to scientific terms not merely by the phenomena being directly studies, but by the structure of theory in which the interpretation of these phenomena is embedded. We may argue that in an ideal world there would be an isomorphism between the languages of psychology and of biochemistry, though even this may be doubted. What is sure is that, at present, such an epistemological isomorphism does not exist and to attempt to impose it artifically is to run the risk of emptying both psychology and biochemistry of their own scientific logics (see earlier chapters).

This is perhaps why the search for correspondents in the field of learning and memory, which are less ambiguous terms than 'aggression', is conceptually and methodologically somewhat easier than for those of other aspects of behaviour and experience. When an animal learns, its subsequent behaviour changes, as a result of experience, in a predictable way which is a function of the learned experience, and the assessment of which is more accessible to both qualitative and quantitative study than in many other areas of behaviour. It follows from the Axiom that, when memory formation occurs, there must be a lasting change in the corresponding cell biology. This change is imposed by learning, and subserves recall; it is 'the' memory trace or engram (which may turn out to be multiple and redundant). This is the object of our enquiries.

As I have said, for an animal to learn, at least in ways accessible to experimentation, the experience with which it is presented must be of some importance to its survival, involving many different aspects, such as danger or pain, or the solution to pressing problems of hunger, thirst or sex. By definition, the registration of these aspects of the experience will themselves have cell biological correspondents, and one major task of the neurobiologist of learning and memory is thus to distinguish (in organisms *capable* of memory) between those cell biological events which correspond to learning and memory formation and those which correspond to predisposing or concomitant behaviours and experiences. Motor activity, stress, or sensory stimulation and other transients, which may be *necessary* for the learning to occur, do not in themselves *constitute* the learning. A good many biochemical and pharmacological studies in this area are concerned with exploring the correspondents of such necessary predispositions and concomitants, rather than of engram formation itself — for instance, the biochemical sequelae of the experience of pain when aversive stimuli are involved.

A parallel biochemical argument to this question of behavioural concomitants applies: if the cell biological remodelling of synaptic membranes involves new protein synthesis, enhanced neuronal energy metabolism may be a necessary concomitant. The cell biology of the engram must be distinguished from such concomitants, and we may formulate a corollary to the Axiom:

Corollary The task of the neurobiologist of learning is to identify the necessary, sufficient and specific cell biological correspondents of memory formation.[3]

The criterion of necessity is specified above; that of sufficiency means that our cell biological statement of correspondent biochemical changes must comprise *all* and *only* those necessary to memory formation; that of exclusivity means that they must correspond to the memory, and not to anything else. In so far as 'memory' is a global word for a set different processes, overlapping temporally and possibly spatially as well, there will be multiple sets of biochemical correspondents, with differing time durations and morphology.

In practice, we may feel satisfied if we can identify a proportion of the necessary and sufficient correspondents; the test of specificity remains at present beyond our grasp.

According to the Axiom, long-term memory corresponds to a lasting change in the cell biology of the brain. It is an empirical question whether this must be detectable as a lasting change in the overall biochemistry of any particular brain region. If long-term memory involved a change in the efficacy of a particular set of synapses, this could correspond to lasting differences in concentration, turnover or structure of particular membrane components or transmitter molecules, for instance. But memory without such lasting changes in total quantity or turnover can be envisaged; the synapses concerned might be remodelled or relocated or the spatial relations between particular cells might be recast without any residual changes in quantity or turnover once the remoulding was complete. In a microsense, of course, there *will* be lasting changes in the distribution of molecules across a cellular region — in the spatiotemporal biochemical co-ordinates of the region. But there will be no such changes in a form which is amenable to empirical biochemical (as opposed, perhaps, to morphological) investigation. There will be a relatively brief and transient biochemical sequence of events as remoulding takes place, followed by reversion to a 'control' state. In fact the experimental observations made to date all do refer to such transients, detectable for at most a few hours or days after-training (some of these experiments are discussed in the next chapter) although the question of more lasting changes remains open. The criteria proposed below are independent of whether the biochemically detectable changes persist; they refer to the processes of memory formation, rather than of its longer-term storage.

Six criteria for correspondence

I propose that any cell biological system or aspect of metabolic biochemistry which is to be considered as a candidate for being part of the engram must as a minimum meet the following six criteria, though I do not present this as an exhaustive list.

Criterion 1 The process or metabolite must show neuroanatomically localised changes in level or rate during memory formation.

This criterion is the most general; to meet it is at best to have achieved necessity but not sufficiency nor exclusivity. The argument that this criterion is necessary follows from the Axiom. The Axiom does not speak, however, to the question of neuroanatomical localisation. It is true that memory could be represented by diffuse changes in many cells rather than in specific ones, as was the basis for holographic analogies for information storage, which had a certain vogue in the 1960s. But even these would not imply that *all* regions of the brain, from cerebellum through cortex to the midbrain, were involved. There must be some specificity and localisation of main effects even on the most delocalised model.

The danger of biochemical artefacts discussed above further reduces the power of this criterion, but it becomes the starting point for all subsequent analysis. Several laboratories have made observations on particular biochemical events relevant to this criterion[4] and have found changes which make biochemical sense as involving the new synthesis of certain proteins, their post-translational modification and involvement in synaptic membrane phenomena. Such effects show a widely different range of brain localisation, varying, as perhaps might be expected, with the task, species and experimenter. It is not always clear whether effects in other brain regions have been ruled out, because sometimes only the 'affected' regions have been examined. Nor can it be certain that the changes are not biochemically secondary to other biochemical events. Of course, in no case do the studies claim to be comprehensive for all of a range of *possible* biochemical changes; negative data, as is well known, are rarely published. Yet in the long run, we will need to know what does not change at least as much as we need to know what does. Certainly the range of potential biochemical candidate systems for correspondents to the engram, based on this criterion alone, can only increase in the next few years.

Criterion 2 The time course of the change in biochemistry must match the time course of the specific phase of memory formation of which it is the correspondent.

The necessity of this criterion is plain, yet is hard to meet — not least because of the ambiguities which surround the phenomenology of memory formation at the psychological level, briefly raised above. If an animal is trained on a task, biochemical measures made some hours or even days later, and a change compared to a control 'non learning' animal is found, the implications are obscure unless a detailed time course of the change has been made.

A change which persists after 24 hours may have arisen during long-term memory formation and be a correspondent of that process; it can scarcely be a correspondent of the early phase of memory formation, but unless it is known how long it persists subsequent to the training experience, its relation to long-term memory may also be problematic.

A change which occurs within minutes of the start of training may be one of two things; a correspondent of short-term memory or part of the biochemical mobilisation involved in 'fixing' long-term memory. It will be observed that meeting Criterion 2 requires a resolution of the phenomenology of memory formation discussed above. As many of the experiments germane to this phenomenology are derived from the biochemical interaction and inhibition studies referred to under Criterion 4 below, the two issues are likely to be resolved together, if at all. Indeed much of what is known or argued about by neuropsychologists who study learning, is a consequence rather than a presumption of biochemical experimentation. It is clear that the duration of distinct phases of memory formation differs with the species, task and the exact experimental design. The search for biochemical correspondents must take cognisance of this and work within the constraints given by a task, the behavioural phenomenology of which has been thoroughly explored.

Criterion 3 Stress, motor activity or other necessary but not sufficient predispositions or concomitants of learning must not in themselves and in the absence of the memory formation result in changes in Criterion 1.

This may be one of the hardest criteria to meet. Note that it does not imply that memory formation is possible without the concomitants, but that the concomitants may occur independently of learning. One problem is the uncertainty at the behavioural level about the stages involved in memory formation, already discussed; devising 'non-learning' controls partly depends on the assumption that unless the experimenter observes a measure of learning on the specific test under consideration, then the animal has learned nothing. But this implies that the organism is a sort of *tabula rasa* on which the only meaningful experience is the one imposed by the experimenter.

In some experiments 'yoke controls' have been used in which a control animal experiences the cues and receives identical stimuli (for instance foot-shocks) to those received by the experimental one, but cannot learn an escape route to avoid the shock. But perhaps in such procedures the 'controls' learn that there is *no* escape, and may hence be *more* stressed than the experimental animals? In some procedures, animals are divided into two populations on the basis of success or failure to achieve a criterion of learning, and these two populations are then compared for a particular biochemical measure. Yet the experimenter only has, by definition, *post hoc* knowledge of whether an animal will or will not be successful in the training; biochemical differences may relate to prior endogenous behavioural differences between the populations which cannot be assessed in the experiment. For most purposes one can only assay an animal biochemically once!

Behind these problems, however, there lies another which relates to the point made previously concerning the reification of descriptions of behavioural/experiental states and processes. Devising controls for concomitants of learning implies (i) that we have a clear inventory of what these concomitants are, (ii) that we can in fact place an animal in a situation in which the concomitants occur without the learning, and (iii) that we have a clear model of the process of learning and memory formation itself in terms of initial registration processes, and short- and long-term consolidation, so that the correspondents of these stages in memory formation can be distinguished.

Despite this formidable set of difficulties, we may reasonably adopt an experimental strategy which attempts to show that in a large number of controls in which experimenter-defined learning is absent, the biochemical change does not occur. A parsimonious interpretation would then be favourable to regarding the biochemical event as a correspondent of the memory formation. Perhaps the control of most general applicability is to compare two groups of animals on a particular task, one which is still learning the task (undertrained) and the other which has already learned it thoroughly (overtrained); both groups will perform the same task, but while one group is learning as it performs, the other, presumably, no longer has anything (or not much) left to learn about it.[5]

Another powerful control is one in which the assumption is made that the engram is normally stored bilaterally, in both halves of the brain. Appropriate procedures (such as 'splitting' the brain by cutting the pathways between the two halves) may then enable learning to occur only in a particular region of a single hemisphere, which can then be compared with the non-learning side of the brain.[6] Such designs will control for brain effects which may be expected to be symmetrically distributed (for instance, generalised circulation-borne hormones) and the sequelae of motor activity; they will not control for changes at the microcirculation level (for which there is increasing evidence) or for changes consequent on input into the trained, compared with the untrained, hemisphere — for instance, sensory stimulation.

Criterion 4 If the cellular/biochemical changes of Criterion 1 are inhibited during the period over which memory formation should occur, the memory formation should be inhibited, and vice versa.

If biochemical correspondents are one-for-one (or even one-for-many) it is logically necessary that inhibition of the correspondents should inhibit memory formation, and there is a substantial body of literature which has been based on this experimental design.[7] It is also logically necessary that the reverse should hold; inhibiting memory formation while holding the other conditions constant should inhibit the correspondent. Problems of interpretation of the experiments in the literature revolve around both behavioural and biochemical issues. If the inhibiting agent (drug, antibiotic, electrocortical stimulation, or whatever) interferes with a concomitant of memory formation (for instance, attention or arousal) a spurious correspondence may be generated with the biochemistry. Administration of the inhibitor may also affect the criteria of acquisition, so that what the animal achieves is actually a 'state-dependent' memory, for the expression of which re-administration of the agent is required. And the agent may affect biochemical processes 'upstream' or 'downstream' of what is actually being measured, as in the case of the protein synthesis and amino acids referred to above.

As a strategic tool for exploring the correspondents of memory the use of inhibitors is thus limited. They have added to our information on the time course of memory formation[8], but the light they cast on the exact nature of the biochemical correspondents is limited to a confirmation of involvement of a correspondent demonstrated using other criteria. Every necessary and sufficient correspondent must fit Criterion 4; not all phenomena which fit Criterion 4 are necessary and sufficient correspondents.

There remains scope for using the reverse form of this criterion. If a biochemical process is suspected of being a correspondent, inhibition of memory formation should inhibit the process; an experiment along these lines is discussed in the next chapter.

Criterion 5 Removal of the anatomical locus (or loci) at which the changes of Criterion 1 occur, should interfere with the process of memory formation and/or recall, depending on when, in relation to the training, the region is removed.

If the changes is connectivity which it is surmised constitute the engram, are anatomically localised rather than diffusely spread in a multiple set of networks across entire regions of the brain, then removal of the locus subsequent to training should eliminate the memory. The regions at which the biochemical changes are observed should correspond to those whose anatomical deletion obliterates the memory. Despite the well-known objection to the interpretation of ablation studies — that what they reveal is not the function of the missing region, but of the rest of the brain in the absence of that region — the localisation criterion is logically necessary, granted an otherwise identified potential correspondent. However, the reverse of this criterion does *not* apply. If ablation of a region disrupts memory formation, it does not follow that biochemical analysis must reveal changes occurring there. The same caveat about the interpretation of anatomical ablation studies must apply to the interpretation of the biochemical inhibition studies referred to in Criterion 4.

There are many studies of the effects of ablation on memory formation and recall, but most are not coupled to biochemical data; the only studies germane to this criterion, of which I am aware, are those of Horn and his colleagues in the chick. Having localised the site of elevated RNA synthesis during imprinting to a specific region, they have been able to show that ablation of the region following training prevents the expression of the imprinting response, while ablation prior to training prevents acquisition of the response.[9] However, it remains arguable even here that both the biochemical change and the anatomical locus are involved in some necessary organising responses of recognition, of the stimulus or of the following responses to it, rather than as the site of the engram itself.

Criterion 6 Neurophysiological recording from the locus (or loci) of the changes of Criterion 1 and 5 should detect altered cellular responses during and/or as a consequence of memory formation.

If engram formation results in a changed connectivity of specific cells at a particular locus, it follows that recordings made from these cells during learning should show changes by comparison with the patterns prior to learning or following memory formation. When the animal recalls the memory again then these cells should again show alteration in firing patterns. That is, there should be a coincidence of anatomical, biochemical and physiological correspondents to engram formation and engram utilisation. Again, an experiment to show this is described in the next chapter.

Beyond the six criteria

These six criteria are, I claim, formally necessary, sufficient and perhaps even specific to any cell biological correspondents of the engram. A seventh may be less so. I am not sure, for instance, whether one should expect a relationship between the 'strength' of any memory and the extent of the biochemical change? Might it be that the more 'powerful' the memory, the greater the extent of the modification to the connections which may be involved? There are intuitive arguments why this might be so, and some relevant evidence,[10] but one must be careful of the generalisation.

There is of course a problem about measuring the 'strength' of a memory. Not all forms of learning can be easily quantitated. Indeed, even a linear preference score need not imply a linear set of internal events. Setting an arbitrary criterion for learning, and ranking performance to criterion as in maze learning or shuttle-box tasks, may give the appearance of being metric to what is actually an ordinal scale. The biochemical and behavioural metrics are constructed in different languages and on different assumptions, and cannot automatically be assumed to be superposable in all details. Another apparent metric results from the assumption that when a batch of animals is trained on all-or-non-learning task, the fraction of the batch which shows recall on test may be equated with a fractional learning score for any individual animal. Such assumptions of quasi-metricity may be operationally convenient; they cannot, however be matched directly against some genuinely metric measure for a biochemical correspondent.

The evidence from the literature of the biochemistry of memory, set in the context of these criteria, does enable one to put together a reasonably convincing picture at least of some of the biochemical events involved in engram formation. I will discuss some of these in more detail in the next chapter. For the purposes of this chapter, I don't want to offer an elaborate model for a possible change in connectivity or a series of biochemical processes corresponding to memory formation. The semi-theoretical literature and the reviews of the field of the last two decades are replete with such models.

There is a multitude of potential ways in which we can envisage synaptic connectivity being altered, and until there are experiments which can distinguish between these possibilities, further speculation is simply burdensome; vacuous for the theoretician, useless to the experimenter. It is enough to argue that the mechanisms of changing connectivity involved in memory formation (if so they be) are likely to conform to general cell-biological and neurochemical principles and not to involve fundamentally new mechanisms. The credibility of potential correspondents must be judged against this background.

Finally, it still seems reasonable, at least as a heuristic, to argue that the biochemical correspondents of engram formation will be found to be similar in form whatever the engram concerned. The specificity of a memory seems more likely to be conferred by the spatial coordinates of the biochemical correspondents — that is, the addresses and connectivity of the cells concerned — rather than molecular differences between the correspondents of different memories. Despite the flurry of enthusiasm for the idea in the 1960s, there seems no reason to postulate 'memory molecules' — molecular, rather than cellular engram formation. However, granted the degree of molecular specificity that is known to exist among different systems of neurons, we may in due course discover that there are differences in detail among the correspondents of different classes of memory. A clarification of the questions involved in the study of the biochemistry of memory formation may help to set our experimental programme onto a sound footing.

Notes and References

1. See discussion by Ruth Hubbard in S. Rose (ed.) *Towards a Liberatory Biology*, Allison & Busby, 1982.
2. V. Reynolds, *The Biology of Human Action*, Freeman, 1976.
3. I owe this formulation to discussions with Patrick Bateson who, with Gabriel Horn, collaborated with me over the period 1969–76 in our initial studies of the biochemistry of imprinting in the chick. See for example, P.P.G. Bateson, 'Neural consequences of early experience in birds', in P. Spencer-Davies (ed.) *Perspectives in Experimental Biology*, Pergamon, 1976, vol. 1, pp. 411–15.
4. There have been many such reviews and symposia over the last decade. See for example, B.W. Agranoff, H.R. Burrell, L.A. Dokas and A.D. Springer, 'Progress in biochemical approaches to learning and memory', in M. Lipton, A. Di Rascio and K. Killam (eds.) *Psychopharmacology*, Raven, 1976; L.R. Squire and N. Butters, (eds.) *Neuropsychology of Memory*, Guildford Press, 1984; and H.J. Matthies (ed.) *Learning and Memory*, Pergamon, 1986.
5. P.P.G. Bateson, 'Are they really the products of learning?', in G. Horn and H. Hinde, (eds.), *Short-term Changes in Neural Activity and Behaviour*, Cambridge, U.P. 1970, pp. 553–64.
6. For reviews see G. Horn, S.P.R. Rose and P.P.G. Bateson, 'Experience and plasticity in the nervous system', *Science*, 181, 1973, pp. 506–14.
7. See for instance L.R. Squire and H.P. Davis, 'The pharmacology of memory: a neurobiological perspective', *Annual Review of Pharmacology*, 21, 1981, pp. 321–56.
8. J.L. McGaugh, 'Time dependent processes in memory storage', *Science*, 153, 1966, pp. 1351–8.
9. G. Horn *Memory, Imprinting and the Brain*, Oxford, 1986.
10. P.P.G. Bateson, G. Horn and S.P.R. Rose, 'Imprinting: correlations between behaviour and incorporation of ^{14}C-uracil into chick brain', *Brain Research*, 84, 1975, pp. 207–20.

10 The Cellular Structure of Memory

In the previous chapter, I discussed how one might approach a rational cell biology of memory formation. I said very little there about the actual detail of experiments or the sort of information that they might yield, or indeed about what cell-biological form an actual engram might take. In this final chapter I want to discuss my own experimental work of the past few years in order to show one approach to this question. It is an approach which I hope combines a respect for reductionist methodology with a commitment to a non-reductionist, dialectical (or integrationist) philosophical standpoint.

Before discussing these experiments, though, a little more background is needed. First, if memories are stored in the brain in terms of changed structure and connections, what form might these changes take?

Back in the 1940s, the Canadian psychologist Donald Hebb produced a model for how learning might occur which has remained the basis of all subsequent theories.[1] For Hebb, the key lay in the synaptic junctions between two nerve cells. He speculated that some of the junctions might begin by being relatively weak or inactive. If circumstances arose in which both cells became active and fired at more-or-less the same time, then the synapses between them might become strengthened, making it more likely that in the future, if one cell fired, then the other would be excited and fire too. For instance, one cell might lie along a pathway in the visual system, the other in that involving taste. An animal seeing an apparently tasty object would bite it. Seeing the object would involve the first cell, tasting it the second; the synapse between the two cells would become strengthened, so that later, when the object was seen again, the firing of the visual cell would tend to make the taste cell fire too, thus invoking the memory of the taste.

With a variety of refinements, the idea of modifiable, or Hebb synapses, remains the basis for most of today's ideas about and experimental approaches to the way that learning occurs and memories are stored. (Though it is only fair to add that there *is* an alternative viewpoint, one which argues that memories are not localisable at all in the brain, but are stored in a distributed form, rather as a holograph stores photographic information. Because my type of experimental approach, or indeed that of cell biology in general, cannot directly distinguish between these possibilities, I don't propose to get involved in that debate here.)

My question is, experimentally, what happens when memories are formed? Neurobiologists began to explore this question in the 1960s, in the first flush of enthusiasm of the then new techniques of molecular biology. The approach was to take over the models for studying animal learning which the psychologists had developed — notably the use of training boxes in which rats or mice had to learn to press a bar to receive food, or to jump to a small shelf when a light flashed to avoid being shocked — and to ask what types of biochemical change took place during such learning. Very soon reports began to come in of big changes in the synthesis of RNA or protein during learning. Further, if the animals were injected with a drug which inhibited protein synthesis just before they were trained it seemed as if they could learn the novel task but not recall it when tested some hours later; hence, it was argued, protein synthesis was necessary for long-term but not short-term memory.

It seemed as if the problem of memory storage would soon be solved and for a few years the field became very fashionable. Then the type of problems I discussed in the last chapter began to appear. The more sensational experiments — such as those claimed to involve 'transfer' of learning from a trained to a naïve animal by injecting RNA extracts — couldn't be replicated. Many of the claims that training produced increases in protein synthesis did turn out to be valid, but only because when an animal is trained, say to avoid footshock, it is also inevitably stressed, and the biochemical changes seemed to be produced by the stress itself rather than 'purely' by memory formation. Hopes for a quick experimental answer faded and many discouraged researchers moved on to other fields.

It is only within the last few years that new approaches have begun to break through the impasse of the 1960s, and it begins to look as if the cellular analysis of learning and memory has at last come of age. In the last chapter, I pointed to three problems which had hindered research on the cell biology of memory, and reserved discussion of the fourth. The time has now come to discuss it. It is one common to all biological research, that of finding the right organism to study. The trouble with working on the cell biology of learning and memory in the rats and mice beloved of psychologists is partly that they have very complex nervous systems containing vast numbers of nerve cells; and partly that the tasks that classical psychology has chosen to teach its experimental subjects are themselves complex, involving subtle skills in already mature animals. Therefore it may well be the case that the actual cellular changes that occur in such forms of learning are very small by comparison with the general 'background' of ongoing biochemical activity.

Three types of model system have begun to prove particularly powerful over the last few years. The first, developed in the USA by Eric Kandel and his colleagues in New York[2] and Dan Alkon at Woods Hole,[3] has been to abandon vertebrates entirely in favour of animals with apparently simpler nervous systems, such as large sea molluscs (*Aplysia* and *Hermissenda*). Such animals have 'brains' built of small numbers of large nerve cells but which nonetheless show some of the features of short-term learning also characteristic of vertebrates. In these molluscs, Kandel and Alkon have been able to define precisely the neural circuits involved in some simple responses (such as the gill withdrawal reflex in *Aplysia*), and to study the ionic changes occurring in key synapses in the circuits during learning.

The second approach, pioneered by Tim Bliss at the National Institute for Medical Research in London and exploited by Gary Lynch and his colleagues in Irvine, California,[4] abandons whole organisms entirely in favour of artificially maintained slices of a particular region of the mammalian brain believed to be involved in memory storage processes, the hippocampus. Cells in these slices retain their electrical properties and can be stimulated by applying electrical pulses to the nerve fibres which enter them. Bliss showed that applying pulses of particular frequencies would result in long-lasting changes in the firing properties of the cells in the hippocampal slice. Could such changes be cellular analogues of memory? Lynch and his group and Bliss himself have been able to explore the electrical and biochemical events at hippocampal synapses when this phenomenon, which has been called long-term potentiation, is induced.

My own approach has been somewhat different. I wanted to continue working with a recognisable form of learning in a vertebrate, but to choose one in which the cellular changes were likely to be large and readily studied. Very young animals have particular merits from this point of view, and in particular young precocial birds, hatched with relatively fully-formed brains, and needing within a few hours to learn a great deal about the world about them if they are to survive in it. The young chick, for instance, must rapidly learn who its mother is so as to follow and take shelter with her (imprinting); and to explore its environment for food by pecking, learning to recognise edible objects from small pebbles or faeces.

Late in the 1960s, in collaboration with Pat Bateson and Gabriel Horn in Cambridge (respectively an ethologist and a neuroanatomist), I began to study the biochemical changes which occurred in the brain during and after imprinting.[5] Later in the 1970s, I switched to the more straightforward model offered by the chick's pecking behaviour. If you offer a chick a small bright object, like a chrome bead, It will peck spontaneously at it within a few seconds; if you make the bead taste unpleasant (for instance, by dipping it into the bitter-tasting but harmless substance methylanthranilate), the chick will peck once, show its disgust by shaking its head vigorously and wiping its beak on the ground, and avoid pecking a similar but dry bead if offered any time up to several days subsequently.

This is called one trial, passive avoidance learning. It is simple, reliable and quick. More than 80 per cent of day-old chicks learn to avoid the bead after a single, 10-second trial. It produces profound changes in the cellular properties of particular regions of the brain — changes in synaptic structure, which can be demonstrated with the electron microscope; in biochemistry, which can be shown using radioactive or immunological markers; and in physiology, which can be shown by recording from the brain of the chicks after training.

In the last chapter, I defined a set of criteria which, I argued, any type of cell biological change had to meet if it were to be regarded as a correspondent of memory formation, that is, an engram. Here, I want to consider my own experiments[6] in the context of these criteria. In brief, the five I want to discuss are that any candidate correspondent must:

(1) be localised to specific regions of the brain;
(2) change in rate or level over a time course compatible with the time course of long-term memory formation;
(3) not be produced by stress, motor or sensory activity or any other necessary but not sufficient predisposition for learning alone, in the absence of learning itself;

Further,

(4) If the cellular change is prevented from occurring, for instance by a drug, then memory formation should not occur, while if memory formation is inhibited, then the cellular change should also be inhibited;
(5) If neurophysiological recordings are made from the site of the cellular change, altered responses should be detected during and/or as a consequence of memory formation.

The first task then, was to discover where in the chick brain (if indeed anywhere!) measurable changes occur as a result of being trained to avoid pecking at a bitter-tasting bead. To discover this requires that we make no prior assumptions about what type of long-term biochemical change might be occurring, but only that we assume that *all* changes which are going to involve cellular reorganisation or the increased synthesis of macromolecules are going to make extra energetic demands on the cell, and hence to increase the uptake and utilisation of the brain's universal source of energy, glucose. A standard neurochemical technique for looking at regional glucose utilisation is to make use of a metabolic analogue of glucose, 2-deoxyglucose (2-DG), which like glucose is taken up into cells and phosphorylated, but cannot be metabolised further. If radioactive 2-DG is used, radioactive 2-deoxyglucose 6-phosphate accumulates in the cells. The amount present can be measured, and is assumed to be proportional to the rate of glucose utilisation by the cells.

We could therefore take two groups of chicks, one group trained by pecking on a water-coated bead (chicks trained like this will go on pecking at a dry bead later) and one on the bitter, methylanthranilate-coated bead (these chicks will avoid a dry bead later, having learned that it is unpleasant). Comparing events in the brains of birds from the two groups should enable us to distinguish the changes associated with learning *not* to peck at the bitter bead.

Each bird is injected with radioactive 2-DG just before or just after training.

Half an hour later they are tested for memory of the bead, and serial microscope sections made of their forebrains. Autoradiograms of these sections then show the amount of radioactivity accumulated in specific brain regions; the autoradiograms can be automatically scanned and a computer plot of optical densities made. Comparing scans from the brains of the methylanthranilate and the water-trained birds showed that there was increased radioactivity, and hence glucose utilisation, in three specific regions, which chick neuroanatomists had previously called medial hyperstriatum ventrale (MHV), paleostriatum augmentatum (PA) and lobus parolfactorius (LPO).

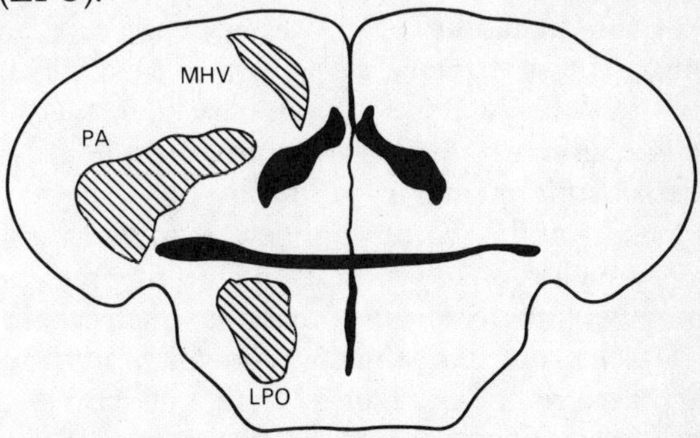

Sketch of section through chick brain showing the regions in the left hemisphere whose cells change their biochemical, physiological and morphological properties when the chick learns to avoid a bitter-tasting bead

Not much is known about the neuroanatomy of these regions, though the MHV in the chick is roughly comparable to parts of the cortex in mammals, and the PA and LPO to regions associated with motor responses; the MHV is also the same region which Gabriel Horn, in Cambridge, has shown to be involved in imprinting in the chick. The experiment meant we could infer that these were sites likely to be involved in some way in responding to the experience of pecking the bitter bead. It was also particularly exciting to find that the metabolic changes tended to be most pronounced in the left, rather than the right, hemisphere MHV and LPO, as we already had behavioural reasons to suspect that there might be some lateralisation of the memory trace, based on some experiments done by Lesley Rogers, at Monash in Australia.

The 2-DG experiment was addressed to the first of my criteria above, as it shows merely that there indeed *are* localised metabolic changes in specific brain regions during training. What could such general changes mean in more specific biochemical terms? My second criterion speaks of the time course of memory formation. When we began these experiments, it was already known, from the work of Marie Gibbs and her colleagues at La Trobe in Australia, that memory formation in the chick for the passive avoidance task occurs in a series of phases over a period of some hours. The first minutes after training form a labile, short-term period, where memory is easily disrupted, and which may involve more than one ionic and biochemical mechanism. Over the first hour or two after training, the long-term memory trace builds up, however, and beyond this time is 'fixed' and cannot be disrupted.

Quite early on in the experimental programme, Marie Gibbs and I were able to show that during the early, labile phases of memory there were biochemical changes involving the acetylcholine system, one of the major neurotransmitters of the brain. In particular, there were transient increases in the activity of the receptor molecule for acetylcholine, present in the post-synaptic membrane. Such increases presumably alter the effect that the transmitter has on the excitability of the post-synaptic cell. But it is with the longer-term changes, those that take place over the first few hours after training and then persist, with which I have been most concerned. Because lasting biochemical modifications of synaptic structures must involve changes in the properties of the synaptic membrane — if only by changing its dimensions — we began by looking at the synthesis of new membrane proteins.

It is easy to study such protein synthesis by injecting radioactively labelled precursors of proteins (amino acids) as markers. Such labelled amino acids can be injected into chicks either before or following training on the water- or the methylanthranilate-coated bead, and the rate of synthesis of proteins in brain regions of interest compared in the two groups. When regions containing the MHV were dissected, and protein radioactivity measured, there was greater radioactivity in proteins from methylanthranilate-trained birds than in the controls trained on water. Purifying the proteins involved revealed that much of the increase was in a particular protein, tubulin, the major constituent of the cellular structures called microtubules. One of the main functions of tubulin in nerve cells lies in organising the traffic of information and molecules between nerve cell bodies, where proteins are synthesised, and the synaptic junctions which may be many microns distant.

But tubulin is a fairly ubiquitous molecule, and it was of less interest to us than the proteins of the synaptic membrane itself. In the membrane, many of the proteins are present in complex forms, notably as phosphoproteins (containing phosphorylated amino acids) and glycoproteins (containing terminal sequences of sugar molecules). The phosphoproteins are seemingly involved in opening and closing ion channels across the membranes, whilst the glycoproteins are concerned with recognition processes whereby, for instance, pre-synaptic and post-synaptic cells contact one another. A good radioactive precursor for glycoproteins is the sugar molecule fucose, and we were soon able to show that training on methylanthranilate also increased fucose incorporation into glycoproteins of the synaptic membranes for at least 24 hours after training. The biochemical mechanism involved in such increases was clarified when we found that training also increased the activity of one of the rate-limiting enzymes involved in activating the fucose molecule for incorporation into glycoproteins, fucokinase. Separating the radioactive glycoproteins by gel electrophoresis showed us that only a limited number of different molecular types, whose molecular weights we know, though not yet their membrane functions, are involved in the response to training.

What might such molecular changes mean to the synapse? The simplest idea would be that they might actually be changing the structure of the synapse in such a way as to make it more-or-less effective in exciting the post-synaptic cell; after all, this is what the Hebb model would predict. But could such changes actually be measured? My colleague Mike Stewart and I, soon joined by other collaborators, notably Andras Csillag from Budapest, began to develop methods for collecting minute samples of tissue from MHV, LPO and PA, making electron micrograph pictures of them, and directly measuring synaptic dimensions. Arduous work, only made possible by new computer technology, which is also able to perform the mathematical transformations of converting measurements made on two dimensional photographs into estimates of three dimensional shapes and volumes.

To our pleasure (and I must confess, somewhat to my surprise!), we found that 24 hours after training on the bead, there are indeed measurable differences in the synaptic structures of the MHV and LPO of methylanthranilate-compared to water-trained chicks. In particular, there are changes in the dimensions of the apposition zone — the region of direct contact between pre- and post-synaptic cells. But the most striking, and to me wholly unexpected, change is a massive increase, of the order of 60 per cent, in the number of synaptic vesicles present per synapse in the left hemisphere MHV and LPO of the methylanthranilate-trained birds. Synaptic vesicles are the 'packages' of acetylcholine or other neurotransmitter within the synapse; and it was as if the training was altering the number of these packages, not just immediately, but as long as 24 hours after a single peck at a bitter bead. As we have available monoclonal antibody markers which recognise specific proteins present within the synaptic vesicles, it now becomes possible to directly correlate these morphological observations with changes in the vesicle proteins.

So far, I've rather deliberately referred to these changes, in biochemistry and cell structure, as the consequences of *training*. To show that they are not just the non-specific concomitants of the training experience — perhaps responses of the brain to the taste of the methylanthranilate, or to the general stress of the unpleasant experience of pecking at a bitter bead, requires experiments directed towards my Criteria 3 and 4. How could such experiments be done? Clearly, one cannot train a bird on a bitter-tasting bead without the taste and the stress occurring, but perhaps one could find a way of the taste and stress occurring without the learning?

One way of getting close to doing this makes use of the fact, which I mentioned in general terms in the last chapter, that electric shock can wipe out short-term memory. If chicks are trained on the bitter bead, and immediately afterwards given a brief, mild electric shock across the head, they appear to lose all memory for the taste of the bead, and when offered the dry bead later they will peck enthusiastically at it just as if they had never tasted the methylanthranilate — that is, they are *amnestic*. Shocking the birds which have pecked at the water-coated bead has no effect on their later pecking behaviour. But the shock only makes the methylanthranilate birds amnestic if it is given within a minute or so of their learning; if it is delayed by, say, ten minutes, then the birds are no longer amnestic and, subsequently avoid pecking a dry bead when offered it as if they had never been shocked.

This effect makes possible an experiment in which two groups of birds are trained on methylanthranilate and subsequently shocked, but one group gets immediate shock, the other delayed shock. So both groups have had essentially identical experiences; that is, they have all pecked at the bitter bead and been given electric shock, but one group later remembers the taste of the bead, while the other is amnestic. If on the biochemical changes I have described are merely the result of the experience of the taste of the methylanthranilate and the associated stress, then there should be no difference between the two groups, because each has had this experience. If the other hand, the biochemical change is something to do with the memory for the taste, then it should occur in the delayed-shock group, which remembers, but not in the immediate-shock group, which forgets.

The full experiment in fact involved six groups of birds, not just the two groups mentioned above, but also similarly shocked groups of birds which had pecked water-coated beads, and unshocked methylanthranilate- and water-trained birds. An hour after training, I tested the birds to check that they did indeed either show recall or were amnestic, then injected radioactive fucose. Three hours later, the amount of fucose which had been incorporated into the membrane glycoproteins was determined.

The results were unequivocal. There was no effect of the electric shocks on fucose incorporation into the water-trained birds. Incorporation in the immediate-shocked, amnestic methylanthranilate birds was the same as into the water groups, but incorporation into the delayed-shocked, non-amnestic birds was as high as in the birds which had been trained but not shocked. Thus the increased production of membrane glycoproteins was not simply the result of the experience of pecking the bitter bead, but only took place if the birds remembered that experience.

If the synthesis of membrane glycoproteins is *necessary* for memory formation to occur, it follows (as in Criterion 4) that if one could prevent that synthesis from taking place, then chicks should be able to learn but not remember the task. We have also now shown this to be the case. The sugar fucose is, chemically, 6-deoxygalactose. In the presence of an analogue of fucose, 2-deoxgalactose, the synthesis of glycoproteins from fucose is prevented. If the 2-deoxygalactose is injected into the chicks just before or just after training, then the birds are amnestic, and peck the bead when tested an hour or more later, just as if they had been given electric shock. However the 2-deoxygalactose is only effective as an amnestic agent if injected within two hours of training — the period during which long-term memory formation is occurring. After this time, preventing glycoprotein synthesis can no longer produce the amnesia.

The implication of the argument and experiments so far is that when birds learn to avoid pecking at a bitter-tasting bead, there is a coordinated series of biochemical changes in the nerve cells of three particular regions of the brain, MHV, LPO and PA. The short-term changes involve activation of the acetylcholine transmitter system, but within the first hours after training, protein and glycoprotein synthetic machinery is mobilised, new glycoproteins are synthesised in the cell bodies and transported to the pre- and post-synaptic sides of the junctions between nerve cells. There, they are incorporated into the membranes of the synapses and the synaptic vesicles, changing their structure and altering their size, presumably in order to strengthen some synapses, or perhaps to weaken others.

All this is at least in accord with a model of memory formation by the modification of Hebb-type synapses, though of course it doesn't *prove* the theory, for there are doubtless other models (for instance, distributive ones) which could fit the data. To go further than this we would have to know fully the 'wiring diagram' of the circuits involving the modifiable synapses we have observed, and that we cannot yet do in any vertebrate, with its complex brain. Such wiring diagrams *have* been traced with some precision in the molluscan preparations I described earlier. What we have shown, at least, is that the biochemical sequence involving synaptic glycoprotein synthesis is both *necessary* and *specific* for the memory formation, though not yet that it is *sufficient* — that is, that other processes are not also involved.

But if we have shown a change in the biochemistry and the structure of synapses, ought it not also to follow that, if these changes have functional meaning, then the electrical properties of the synapses are also modified? This was the point of my fifth criterion. And recently Roger Mason, a graduate student in the research group, and I have been able to test this possibility too. The electrical properties of groups of cells can be explored in anaesthetised chicks by inserting a recording electrode into specific regions of the brain and measuring the spontaneous firing rate of the cells in that region. Knowing that the biochemical and morphological changes involved the MHV gave us the obvious place to look for the neurophysiological changes too (though we have of course checked that changes do not occur in other, control regions).

I trained birds on either methylanthranilate or water, and gave them to Roger Mason for the recording. He was not told which were the methylanthranilate and which the water birds till the entire experimental series was complete. By the time he had recorded from sixteen birds, he told me that he thought he could assign them to the two groups even without me giving him the code, and he was right in fourteen out of the sixteen! Full statistical analysis shows that in the MHV of methylanthranilate-trained birds, compared with the water controls, there is a massive increase — of up to fourfold — in the rate of 'bursts' of high-frequency firing of the nerve cells. This increased bursting activity persists for many hours after training, and is abolished if the birds are made amnestic; as with the glycoprotein synthesis, we therefore have good reason to believe it is part of the process of memory formation.

This is the first time a phenomenon of this type has been observed following a training experience, and what makes it particularly interesting is that these 'bursting' responses seem to be similar in a number of ways to what happens during long-term potentiation in the mammalian hippocampus, the use of which as a model system for memory I mentioned earlier. Could we be getting closer to a general mechanism for synaptic modulation in memory?

There's a very long way to go yet. General theories apart, even in the chick we need to be much clearer about the biochemical nature of the glycoproteins, the possible involvement of phosphorylated membrane proteins and the exact transmitters involved. Concentrating on pre-synaptic events should not allow us to exclude involvement of cellular changes on the post-synaptic side too. To understand the neurophysiology, we need to be able to record not merely from the anaesthetised chick, but to discover what types of electrical changes might be occurring in the awake bird during training and the subsequent period of memory formation. Above all we need to know a great deal more about the significance of our three regions of change, the MHV, LPO and PA, in the general economy and functioning of the brain as a harmonious integrated whole (Criterion 6 above). What, for instance, is implied by the fact that some of the changes occur in these regions in both hemispheres, while others are confined to the left?

The approach to the analysis of the cell biology of memory formation I have described here involves the deliberate application of a reductionist research strategy to a phenomenon, memory; the ultimate understnding of which cannot, however be achieved except in the context of an integrative, dialectical concept of the organism and the part played by memory in the continuity of its existence. My approach also deliberately challenges that great divide in biology, psychology and philosophy between descriptions of events and phenomena at different levels, those of mind and brain, of behaviour and of molecular mechanisms. I am not arguing that memory, even so simple a memory as a chick's avoiding a bead it has once experienced as tasting bitter, can be reduced to molecules. Indeed, if our model is correct, the molecular changes that we observe are in one sense relatively trivial 'housekeeping' processes for cells whose form and connectivity are being remodelled during the learning. The memory does not lie in the molecules at all, but in some sense in the reorganised cellular networks that the molecules form.

Our experimental difficulties in approaching this problem are compounded by the theoretical complexities of the need to use such reductionist strategies to understand a phenomenon which itself cannot be so reduced. Despite this, the prospect of understanding the cellular processes involved in memory now looks bright; we have the models, the methods and the beginnings of some experimental data too. The next few years could bring some real understanding.

Whether we would then be able to convince more radical critics of any type of even methodological reductionism that we knew something worthwhile about the brain mechanisms of memory remains to be seen. A hundred and fifty years ago, Jane Austen put the central question which informs my research and theorising into the mouth of the stoic heroine of *Mansfield Park*, Fanny Price. There is no more fitting ending to this chapter:

> If any one faculty of our nature may be called *more* wonderful than the rest, I do think it is memory. There seems something more speakingly incomprehensible in the powers, the failures, the inequalities of memory, than in any other of our intelligences. The memory is sometimes so retentive, so serviceable, so obedient — at others, so bewildered and so weak — and at others again, so tyrannic, so beyond controul! — We are to be sure a miracle in every way — but our powers of recollecting and forgetting, do seem peculiarly past finding out.[7]

Notes and references

1. D.O. Hebb, *The Organisation of Behavior*, Wiley, 1949.
2. E.R. Kandel and J.H. Schwartz in P.H. Abelson, E. Butz and S.H. Snyder (eds.) *Neuroscience, AAAS Pub* 84–13, 1985, pp. 381–402.
3. D.L. Alkon in D.L. Alkon and C.D. Woody (eds.), *Neural Mechanisms of Conditioning*, Plenum, 1986, pp. 3–18
4. G. Lynch and M. Baudry in L.R. Squire and N. Butters (eds.), *Neuropsychology of Memory*, Guildford, 1985, pp. 513–20.
5. The imprinting research has been reviewed in, for example, G. Horn, S.P.R. Rose and P.P.G. Bateson, 'Experience and plasticity in the nervous system', *Science*, 181, 1973, pp. 506–14; and most recently in Gabriel Horn's book *Memory, Imprinting and the Brain*, Oxford, 1986.
6. The key papers which report the research findings discussed in this chapter are:-

   M. Kossut and S.P.R. Rose, 'Differential 2-deoxyglucose uptake into chick brain structures during passive avoidance training', *Neuroscience*, 12, 1984, pp. 971–7.

   S.P.R. Rose, M. Gibbs and J. Hambley, 'Transient increases in forebrain muscarinic cholinergic receptors following passive avoidance learning', *Neuroscience*, 5, 1980, pp. 69–72.

   R. Mileusnic, S.P.R. Rose and P. Tillson, 'Passive avoidance learning results in region-specific changes in concentrations of and incorporation into colchicine-binding proteins in the chick brain', *Journal of Neurochemistry*, 34, 1980, pp. 1007–14.

   R.D. Burgoyne and S.P.R. Rose, 'Subcellular localisation of increased incorporation of ^{3}H-fucose following passive avoidance learning in the chick', *Neuroscience Letters*, 19, 1980, pp. 343–348.

   M.G. Stewart, S.P.R. Rose, T.S. King, P.L.A. Gabbott and R. Bourne, 'Hemispheric asymmetry of synapses in the chick medial hyperstriatum ventrale following passive avoidance training: a stereological investigation', *Developmental Brain Research*, 12, 1984, pp. 261–9.

   S.P.R. Rose and S. Harding, 'Training increases ^{3}H-fucose incorporation into chick brain only if followed by memory storage', *Neuroscience*, 12, 1984, pp. 663–7.

   S.P.R. Rose and R. Jork, 'long term memory formation in chicks is blocked by 2-deoxygalactose, a fucose analogue', *Behavioural and Neural Biology*, in press, 1987.

   R.G. Mason and S.P.R. Rose 'lasting changes in spontaneous multi-unit activity in the chick brain following massive avoidance training', *Neuroscience*, in press, 1987.
7. Jane Austen, *Mansfield Park*, Penguin edition, p. 222.

Index